THE SOFTWARE VULNERABILITY GUIDE

LIMITED WARRANTY AND DISCLAIMER OF LIABILITY

THE CD-ROM THAT ACCOMPANIES THE BOOK MAY BE USED ON A SINGLE PC ONLY. THE LICENSE DOES NOT PERMIT THE USE ON A NETWORK (OF ANY KIND). YOU FURTHER AGREE THAT THIS LICENSE GRANTS PERMISSION TO USE THE PRODUCTS CONTAINED HEREIN, BUT DOES NOT GIVE YOU RIGHT OF OWNERSHIP TO ANY OF THE CONTENT OR PRODUCT CONTAINED ON THIS CD-ROM. USE OF THIRD-PARTY SOFTWARE CONTAINED ON THIS CD-ROM IS LIMITED TO AND SUBJECT TO LICENSING TERMS FOR THE RESPECTIVE PRODUCTS.

CHARLES RIVER MEDIA, INC. ("CRM") AND/OR ANYONE WHO HAS BEEN INVOLVED IN THE WRITING, CREATION, OR PRODUCTION OF THE ACCOMPANYING CODE ("THE SOFTWARE") OR THE THIRD-PARTY PRODUCTS CONTAINED ON THE CD-ROM OR TEXTUAL MATERIAL IN THE BOOK, CANNOT AND DO NOT WARRANT THE PERFORMANCE OR RESULTS THAT MAY BE OBTAINED BY USING THE SOFTWARE OR CONTENTS OF THE BOOK. THE AUTHOR AND PUBLISHER HAVE USED THEIR BEST EFFORTS TO ENSURE THE ACCURACY AND FUNCTIONALITY OF THE TEXTUAL MATERIAL AND PROGRAMS CONTAINED HEREIN. WE HOWEVER, MAKE NO WARRANTY OF ANY KIND, EXPRESS OR IMPLIED, REGARDING THE PERFORMANCE OF THESE PROGRAMS OR CONTENTS. THE SOFTWARE IS SOLD "AS IS" WITHOUT WARRANTY (EXCEPT FOR DEFECTIVE MATERIALS USED IN MANUFACTURING THE DISK OR DUE TO FAULTY WORKMANSHIP).

THE AUTHOR, THE PUBLISHER, DEVELOPERS OF THIRD-PARTY SOFTWARE, AND ANYONE INVOLVED IN THE PRODUCTION AND MANUFACTURING OF THIS WORK SHALL NOT BE LIABLE FOR DAMAGES OF ANY KIND ARISING OUT OF THE USE OF (OR THE INABILITY TO USE) THE PROGRAMS, SOURCE CODE, OR TEXTUAL MATERIAL CONTAINED IN THIS PUBLICATION. THIS INCLUDES, BUT IS NOT LIMITED TO, LOSS OF REVENUE OR PROFIT, OR OTHER INCIDENTAL OR CONSEQUENTIAL DAMAGES ARISING OUT OF THE USE OF THE PRODUCT.

THE SOLE REMEDY IN THE EVENT OF A CLAIM OF ANY KIND IS EXPRESSLY LIMITED TO REPLACEMENT OF THE BOOK AND/OR CD-ROM, AND ONLY AT THE DISCRETION OF CRM.

THE USE OF "IMPLIED WARRANTY" AND CERTAIN "EXCLUSIONS" VARIES FROM STATE TO STATE, AND MAY NOT APPLY TO THE PURCHASER OF THIS PRODUCT.

THE SOFTWARE VULNERABILITY GUIDE

HERBERT H. THOMPSON

SCOTT G. CHASE

CHARLES RIVER MEDIA, INC.
Hingham, Massachusetts

Copyright 2005 by CHARLES RIVER MEDIA, INC.
All rights reserved.

No part of this publication may be reproduced in any way, stored in a retrieval system of any type, or transmitted by any means or media, electronic or mechanical, including, but not limited to, photocopy, recording, or scanning, without *prior permission in writing* from the publisher.

Acquisitions Editor: James Walsh
Cover Design: Tyler Creative

CHARLES RIVER MEDIA, INC.
10 Downer Avenue
Hingham, Massachusetts 02043
781-740-0400
781-740-8816 (FAX)
info@charlesriver.com
www.charlesriver.com

This book is printed on acid-free paper.

Herbert H. Thompson and Scott G. Chase. *The Software Vulnerability Guide.*
ISBN: 1-58450-358-0

All brand names and product names mentioned in this book are trademarks or service marks of their respective companies. Any omission or misuse (of any kind) of service marks or trademarks should not be regarded as intent to infringe on the property of others. The publisher recognizes and respects all marks used by companies, manufacturers, and developers as a means to distinguish their products.

Library of Congress Cataloging-in-Publication Data
Thompson, Herbert H.
  The software vulnerability guide / Herbert H. Thompson and Scott G. Chase.— 1st ed.
    p. cm.
  Includes index.
  ISBN 1-58450-358-0 (pbk. with cd-rom : alk. paper)
 1. Computer security. 2. Computer software—Development. 3. Computer networks—Security measures. I. Chase, Scott G. II. Title.
  QA76.9.A25T475 2005
  005.8—dc22
                                    2005010049

Printed in the United States of America
05 7 6 5 4 3 2 First Edition

CHARLES RIVER MEDIA titles are available for site license or bulk purchase by institutions, user groups, corporations, etc. For additional information, please contact the Special Sales Department at 781-740-0400.

Requests for replacement of a defective CD-ROM must be accompanied by the original disc, your mailing address, telephone number, date of purchase, and purchase price. Please state the nature of the problem, and send the information to CHARLES RIVER MEDIA, INC., 10 Downer Avenue, Hingham, Massachusetts 02043. CRM's sole obligation to the purchaser is to replace the disc, based on defective materials or faulty workmanship, but not on the operation or functionality of the product.

Contents

	Acknowledgments	xiii
Part I	Introduction	1
1	**A Call to Action**	3
	Security as a Call to Action for Developers	4
	Why Care about Security	6
	Thinking Differently about Security	8
	Entering the Era of Software Security	9
	Why We Wrote This Book and Why You Should Read It	10
	How This Book Is Structured	13
	Who We Are	17
	References	18
2	**Security Background**	19
	Hacker versus Cracker versus Attacker: The Language of Computer Security	20
	Legal and Ethical Issues Surrounding Computer Security	23
	Federal Laws Related to Illegal Computer Use	23
	Ethical Reporting of Security Vulnerabilities	26
	Networking Basics	26
	Networking References	35
	References	36
3	**Some Useful Tools**	37
	Security Scanners	38
	Comprehensive Scanning Tools	38
	Nmap and Network Scanners	41

Packet Sniffing and Spoofing	42
Hacking and Cracking Tools	44
Password Crackers	45
Packet Generation and Replay	45
Network Fuzzing	46
Web Site Test Tools	47
Reverse Engineering Tools	47
Source and Binary Scanners	48
Specialty Editors	49
API and System Monitors	49
Disassemblers	50
Using Debuggers for Security Testing	50
Commercial Tools	53
Retina	53
AppScan	53
WebProxy	53
Holodeck	53
For More Information	55

Part II System-Level Attacks — 57

4 Problems with Permissions — 59

The Bell-Lapadula Model	60
Description	62
Finding Programs with the Supervisor Bit Set	64
Attacking Supervisor Mode Programs by Finding Side-Effect Functionality	64
Attacking Supervisor Mode Programs by Exploiting a Buffer Overrun	67
Windows: Not Immune From, but Less Prone to, Escalation of Privilege	68
Fixing This Vulnerability	69

		The setuid() and seteuid() System Calls	69
		Summary Sheet—Running with Elevated Privilege	70
		References	71

5 Permitting Default or Weak Passwords — 73

- Finding Default and Weak Passwords — 75
  - Building a Password Cracker — 76
  - Using a Dictionary Helper — 78
  - Writing the Main Crack Routine — 80
  - Putting It Together — 83
- Fixing This Vulnerability — 83
- Summary Sheet—Permitting Default of Weak Passwords — 85
- References — 86

6 Shells, Scripts, and Macros — 87

- Description — 88
  - Embedded Script Languages and Command Interpreters — 89
  - Document Markup — 90
  - JavaScript — 90
  - Safe for Scripting ActiveX Controls — 91
  - Database Stored Procedures — 91
  - Macro Expansion in Logs and Messages — 91
- Fixing This Problem — 92
- Summary Sheet—Shells, Scripts, and Macros — 93
- References — 94

7 Dynamic Linking and Loading — 95

- Finding This Vulnerability — 100
- Fixing This Vulnerability — 101
  - Explicit Linking and Loading of a DLL — 102
- Summary Sheet—Dynamic Linking and Loading — 103
- References — 104

Part III Data Parsing — 105

8 Buffer Overflow Vulnerabilities — 107
- Stack Overflows — 109
- Exploiting Stack Overflows — 113
- Heap Overflows — 116
- Exploiting Buffer Overflows: Beyond the Stack — 122
- Finding This Vulnerability — 127
  - White-Box Testing Techniques and Tools — 128
  - Black-Box Testing Techniques and Tools — 128
- Fixing This Vulnerability — 130
- Summary Sheet—Buffer Overflows — 131
- Endnotes — 132
- References — 132

9 Proprietary Formats and Protocols — 133
- Description — 134
  - Same Data, Many Formats — 135
- Using "Fuzzing" to Find Vulnerabilities in File Formats and Protocols — 138
- Preventing Problems with Proprietary Formats and Protocols — 147
- Summary Sheet—Proprietary Formats and Protocols — 148

10 Format String Vulnerabilities — 151
- The Format Family — 156
- Exploiting Format String Vulnerabilities — 158
- Finding This Vulnerability — 168
  - Fixing This Vulnerability — 169
- Summary Sheet—Format String Vulnerabilities — 170
- References — 171

11 Integer Overflow Vulnerabilities — 173
- Exploiting Integer Overflow Vulnerabilities — 179
- Finding This Vulnerability — 179

	Fixing This Vulnerability	181
	Summary Sheet—Integer Overflows	182
	References	183

Part IV Information Disclosure — 185

12 Storing Passwords in Plain Text — 187

- Finding This Vulnerability — 188
- Fixing This Vulnerability — 196
  - Using the Unix Password Hashing Functions — 197
  - Using `CryptCreateHash` and `CryptHashData` in Windows — 198
- Summary Sheet—Storing Passwords in Plain Text — 198
- References — 200

13 Creating Temporary Files — 201

- Finding This Vulnerability — 206
- Fixing This Vulnerability — 207
- Summary Sheet—Creating Temporary Files — 207
- References — 209

14 Leaving Things in Memory — 211

- Description — 212
  - Finding Exposed Data in Memory — 214
- Fixing This Problem — 221
- Summary Sheet—Leaving Things in Memory — 221
- Endnote — 222
- References — 222

15 The Swap File and Incomplete Deletes — 223

- Using a Disk Editor to Find Confidential Data Fragments — 226
- Fixing This Problem — 230
- Summary Sheet—The Swap File and Incomplete Deletes — 232

Contents

Part V	**On the Wire**	**235**
16	**Spoofing and Man-in-the-Middle Attacks**	**237**
	Finding Spoofing and Man-in-the-Middle Attacks	238
	Connection Hijacking	240
	Name Server Cache Poisoning	247
	Spoofing at the Application Level	250
	Other Kinds of Man-in-the-Middle Attacks: DHCP and 802.11	252
	Preventing Spoofing and Man-in-the-Middle Attacks	252
	Summary Sheet—Spoofing and Man-in-the-Middle Attacks	252
	References	254
17	**Volunteering Too Much Information**	**255**
	Finding This Vulnerability	260
	Fixing This Vulnerability	261
	Summary Sheet—Revealing Too Much Information	263
Part VI	**Web Sites**	**265**
18	**Cross-Site Scripting**	**267**
	Finding Cross-Site Scripting Vulnerabilities	271
	Fixing This Vulnerability	274
	Preventing More Advanced Cross-Site Scripting Vulnerabilities	275
	HTML-Encoding Output	275
	Summary Sheet—Cross-Site Scripting	276
19	**Forceful Browsing**	**277**
	Description	278
	Finding Forceful Browsing Vulnerabilities	281
	Building a Forceful Browsing Test Tool	283
	Preventing Forceful Browsing	295
	Summary Sheet—Forceful Browsing	295

20 Parameter Tampering, Cookie Poisoning, and Hidden Field Manipulation — 297

- Cookie Values — 301
- Form Data — 302
- Query Strings — 306
- HTTP Header Tampering — 306
- Finding This Vulnerability — 307
- Fixing This Vulnerability — 308
- Summary Sheet—Parameter Tampering, Cookie Poisoning, and Hidden Field Manipulation — 309
- References — 310

21 SQL Injection Vulnerabilities — 311

- Exploiting Sites Through SQL Injection — 316
- Finding This Vulnerability — 319
  - Index.html — 320
  - Process.asp — 321
- Fixing This Vulnerability — 322
  - Process.asp — 322
- Summary Sheet—SQL Injection — 323
- References — 324

22 Additional Browser Security Issues — 325

- The Domain Security Model — 326
- Unsafe ActiveX Controls — 328
- Spoofing of URLs in the Browser — 329
- MIME Type Spoofing — 330
- Uncommon URL Schemes — 330
- Browser Helper Objects — 331
- Summary Sheet—Additional Browser Security Issues — 331

Part VII	Conclusion	333
23	Conclusion	335
	Learning from Vulnerabilities	338
	Where to Go Next	338
	References	339
	Appendix A: About the CD-ROM	341
	Appendix B: Open Source Software Licenses	343
	Index	349

Acknowledgments

Many people helped us in the preparation of this book. We would like to express our thanks to the editors and staff at Charles River Media, especially Jim Walsh for his patience during the long process of turning this book from an idea into reality. Michael Howard and Pete Krawczyk provided outstanding input and criticism as technical reviewers.

Richard Ford of Florida Tech helped review the earliest drafts of our ideas. Mike Andrews of Foundstone and Florence Mottay of Security Innovation have provided invaluable insight and advice throughout the project. Matthew Oertle and Michael Cooper at SI Government Solutions helped us test the source code and sample tools we've provided.

James Whittaker, author of *How to Break Software* and co-author of *How to Break Software Security* with Thompson, has been an inspiring teacher and mentor to us since our first project with him in 1998. Without his continued support of us first as students and then as professionals, this project would not have been possible.

The whole team at Security Innovation and SI Government Solutions has been extremely supportive: Ed Adams, Maureen Adams, Scott Bradley, Jason Taylor, Ady Kakrania, Joe Basirco, Jack Whitehouse, Fabien Casteran, Chin Dou, Andres De Vivanco, Nick Laird, Tiffany Cooper, Terry Gillette, Helayne Ray, Jessica Edwards, Brian Shirey, Rusty Wagner, Matt Taylor, Mike Simms, Matt Heine, Chris Baptist, and George Grachis. It is through our daily work with these dedicated security professionals that we have "battle tested" the techniques in this book.

Finally, we would like to thank the people closest to us for being supportive while we wrote this book: Herbert (Dad), Frankie Mae, Maria, Andrea, Patrick, Timothy, Stephen, Robert, Cheryle, Nicole, Adam, and Andrew.

Part I
Introduction

1 A Call to Action

In This Chapter

- Security as a Call to Action for Developers
- Why We Wrote This Book and Why You Should Read It
- How This Book Is Structured
- Who We Are
- References

As the popularity of the Internet has grown, so has the number of individuals who use it to attack others. Most targeted attacks, viruses, and worms have been made possible by vulnerabilities in software that read untrusted data from the network. For years, consumers have ignored the existence of software flaws, and the response to their existence by the IT industry has been the creation of defenses at the perimeter of the network. These defenses have taken the form of firewalls, routers, antivirus software, patch update applications, e-mail filters, and other applications that attempt to protect the user from external attackers. The standard practice has been to draw a line in the sand, with the bad guys on one side and critical information assets on the other. This is the paradigm that has driven the network security market for years, but it makes one fatal assumption: that we can detect and prevent malicious actions at the network level. Despite the onslaught of new products and marketing literature from network security vendors, though, security is not a problem that can be solved completely with better firewalls and antivirus software.

SECURITY AS A CALL TO ACTION FOR DEVELOPERS

The key issue is that data used to exploit these flaws is usually completely indistinguishable from legitimate application data when viewed at the network level and out of context. Consider, for example, the following hexadecimal values:

6A0068B0FB110068D5FB11006A00FF1588204000

If these characters are interpreted as ASCII values (text), we have the following string:

jh°ûhÕûjÿˆ @

If these characters are a part of an image, an audio file, an executable, or a data file, these values can be interpreted as almost anything. The reason our data is not immediately recognizable as *something* is that what it is depends on the context in which it's interpreted. Without exact knowledge of the expected inputs of an application, a firewall can't determine whether this input is intended or unintended, malicious or benign. If the applications themselves cannot determine correct from incorrect input in some cases, how can an outside appliance?

In one context, our code is actually quite malicious. When supplied in exactly the right place, as part of a long string passed to an unchecked buffer on the stack, it can be interpreted as machine instructions that cause a Windows message box to appear within the affected application (see Figure 1.1). Maybe a message box is not that malicious. However, similar sized snippets of machine code can do some very malicious things, including formatting the disk drive or bootstrapping additional functionality.

FIGURE 1.1 A Windows message box created with our shell code.

ANATOMY OF A SHELLCODE

Our code from the first example is really a tiny program, sometimes called a *shellcode*, within the security community. Shellcode is malicious code that is intended to execute inside an exploited application. This particular block of code (see Table 1.1 for an interpretation of the block) launches a Windows message box,

In the strictest sense, this is not "shellcode" because it doesn't launch a shell. The term has evolved, however, to be used interchangeably with hexadecimal machine instructions that are executed through a vulnerability like a buffer overflow.

TABLE 1.1 Shell Code Interpretation

6A 00	Push x00	Parameter describing type of message box.
68 B0 FB 11 00	Push x0011FBB0	Pointer to the message box caption text.
68 D5 FB 11 00	Push x0011FBD5	Pointer to message box body text.
6A 00	Push x00	Handle to a window.
FF 15 88 20 40 00	Call User32.MessageBoxA	Calling the Windows message box function. In this case we are calling indirectly through a pointer.

Without specific knowledge of the buffer overflow vulnerability and in-depth knowledge of the state of the application reading this data, a firewall has no hope of protecting the vulnerable system. The same holds true of many other add-on software protections that have the near impossible task of generically patching against flaws in software.

Our only hope then is to address vulnerabilities at their root: in software. Security is a problem that needs to be wholeheartedly addressed by software developers and testers and it is a problem that must be addressed by the development organization and not just pushed to operations. But, obstacles exist. The software development community has seen a rash of new programming paradigms, methodologies, and development environments, and still the number of security flaws in software has risen substantially over the last several years. Figure 1.2 shows the number of new vulnerabilities reported to CERT from 1999 to 2003.

6 The Software Vulnerability Guide

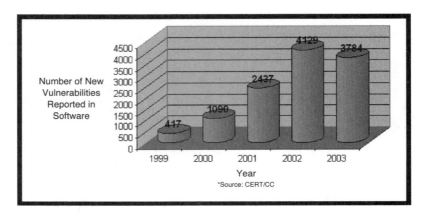

FIGURE 1.2   The number of new vulnerabilities reported to CERT from 1999 to 2003. Data compiled by CERT [Cert04].

The graph in Figure 1.2 shows the number of new vulnerabilities reported in software doubling yearly from 1999 to 2002, and software vulnerabilities were reaching epidemic proportions. In 2002 the biggest names in the commercial software development industry—companies like Microsoft and IBM—shifted the way they were building software in response to the need for security. They began pushing security into all stages of the development lifecycle, and books emerged such as Michael Howard and David LeBlanc's *Writing Secure Code* [Howard02] that finally offered a roadmap for developers to make more secure software. Perhaps the biggest push came from Microsoft, who invested heavily in training for developers and testers and implemented their Trustworthy Computing Security Development Lifecycle (published publicly in December 2004 [Lipner04]). This security push by the large vendors ignited an industry-wide awareness of security that had immediate impact in rapid production environments such as e-commerce sites, thus lowering the number of new vulnerabilities reported. The expectation is that this trend will continue to be felt as old products are replaced with newer versions that have had the benefit of this paradigm shift. Corporations, governments, and vendors have begun to realize that we need to look at software bugs, requirements, and development differently to adapt to the new security-savvy consumer. We wrote this book to help both software developers and testers better understand the underlying security flaws in software and to be an easy reference guide to security bugs in software.

Why Care about Security

For many corporations, security now has a better understood business case. Software vendors are witnessing the emergence of the security-aware consumer, one who makes purchasing decisions not just on initial price and utility, but who also demands proof that vendors have done a reasonable job of security testing their products. Corporations are now starting to consider the "total cost of ownership" of an application instead of just the cost to purchase and deploy it. They are starting to realize that downtime, data theft, and cyber-vandalism made possible by security flaws in software are part of the cost incurred by deploying vulnerable software. Technology analyst firms now advise their clients to demand some proof of security due diligence from their software vendors. What does this mean for development organizations? Certainly no generally accepted certifications or standardized test suites exist to verify an application or a solution as secure. Business consumers, though, are likely to be the first to ask the tough questions. How well versed are your developers on security? How was this product tested? What secure coding practices did you follow? What methods, processes, techniques, outsourcers, and people did you devote to making sure this product isn't riddled with security holes? In competitive software markets, such as databases, how well a vendor is prepared to answer these questions might be the single biggest factor in who wins the contract. Software security is more than a vendor problem, though. Most corporations write a significant amount of software for internal use or for the Web. Now, more than ever, this software needs to be developed with security in mind and tested for resilience to inevitable attack.

VALIDATE ALL INPUT!

Nearly half of all the attacks in this book are the result of malicious user input that has a security side effect. Part III of this book deals with this problem in detail. Poor input validation is the cause of most of the vulnerabilities described in Part VI of this book as well. The classic user input vulnerability, the buffer overflow, results from copying data past the end of a buffer on the stack, which can overwrite the return pointer (the address to which control flow returns when a procedure is exited) with arbitrary data. If this occurs in data that comes from a remote system (such as a request to a server application), an attacker can use this side effect behavior to take control of the application. Consider the following two snippets of C++ code:

```
char buf[BUF_SIZE];
char* input = getRemoteUserInput();
strcpy(buf, input);
```

and

```
char buf[BUF_SIZE];
char* input = getRemoteUserInput();
strncpy(buf, input, BUF_SIZE-1);
```

For user inputs of lengths less than BUF_SIZE-1, the two snippets behave similarly. However, if the user supplies an input that is BUF_SIZE or greater, the two snippets behave differently. strcpy continues copying the user-supplied data until it finds a null character in the input, overwriting whatever data follows buf in memory. strncpy copies only the first BUF_SIZE bytes, truncating the input but preserving the integrity of the stack, diminishing the chances of a buffer overflow. For this reason, strncpy, and bounded copying generally, is preferred over unbounded copying. If this doesn't make sense, don't worry. Part III of this book explains this in great detail. In the mean time, though, stop using strcpy!

Thinking Differently about Security

In a typical development project we move from requirements and specifications to design and code. Requirements tell programmers what an application, component, or function *should* do. They are usually pretty good at describing how component interfaces should work, the type of data (or inputs) these components will receive, the manipulation that should be done on data, and the eventual outputs of a module. Developers are then tasked with writing components, and testers create tests that feed the application data and look for the presence of correct output. Requirements are great to test against because the verification process boils down to translating specifications of what a product is designed to do and then running test cases that check to see if actual behavior matches specified behavior. The problem with security defects is that the focus of the process has been on creating the correct result without focusing on *how* the application produces that result.

It's the "how" and the "what else did the application do" that are important to security. Consider a simple function that accepts a string of five characters and is expected to return a string, also five characters long but with the characters in reverse order. Therefore, if you were to supply the string "12345", you would expect the output to be "54321". These are very simple requirements for a very basic function. It's easy to imagine test cases for this function, a series of strings with varying characters, all with verifiable results. The astute functional tester would certainly try to vary the length of the input by applying strings of zero length up to hundreds or thousands of kilobytes, expecting to receive an error message if the string was not exactly five characters long. Different developers might choose to implement this function in different ways, possibly using arrays, structures, or temporary files. Each implementation of our simple function might be functionally correct and might pass the test cases discussed. Now, imagine that some other, unspecified security concerns were at play. What if this string were a password or an encryption key? In this case, the requirements would undoubtedly be the same, but the implementation that stores the string in a temporary file would be grossly insecure. In this case, for example, the functionally correct option of writing the string out to a temp file obviously wouldn't be a good idea. We see then that there can be a discrepancy between *secure* and *correct*. This is because security is *contextual*—it depends on the totality of circumstances, and this can be different for different applications. It can also be different from what the developer expected. This is why security understanding by developers is so important; no add-on tool or technology can adequately determine this context.

Partly this is due to a lack of education. It is rare that a software developer knows how easy it is to exploit a buffer overflow or that a Web developer understands how to use SQL injection to gain control of a Web server and its data. Yet, a motivated teenager can learn these skills in a matter of days in the back alleys of the information superhighway. It isn't difficult then to understand why attackers are often so successful at breaking applications and breaking into networks. To build applications that are more resistant to attackers, software developers and testers need to understand these techniques. For each vulnerability discussed in this book, we provide some insight into the attacker mindset and the techniques that are used every day to find and exploit software security flaws. To build and deploy secure systems we must know the tools and techniques of our adversaries.

Entering the Era of Software Security

The new focus of consumers on security has forced vendors to commit to producing more secure software. Developers are starting to become more attuned to the security implications of their code. Also we have a new breed of software tester, one who is focused exclusively on security. Organizations are also beginning to realize the limitations of network defenses when it comes to preventing attack. When we have this in mind, we can see that the responsibility to make their products more secure falls squarely on the shoulders of designers, architects, developers, and testers. As corporations realize that firewalls and other perimeter defenses are ultimately an incomplete defense for their organization, they are now starting to ask software vendors the tough questions about security: Do you have a dedicated team to assess and respond to security vulnerability reports in your products? What process improvements have you made as a result of vulnerabilities reported in your software? What training does your development and testing organizations receive on security? What percentage of your test team is focused on security? For many software vendors, the answers to these questions can be concerning.

Security, though, has become a significant market discriminator for software products. The result is a strong push for vendors to fortify their applications and the critical need for information to help developers and testers make the products they ship more secure. However, the development, design, and test communities still have a fundamental lack of knowledge of security. This need is especially acute given that until recently there was no course on secure programming techniques or security testing in any Computer Science curriculum at any university in the United States. Some progress is being made in this area. Professor Daniel J. Bernstein recently taught an experimental vulnerability course at the University of Chicago. The students in this class were told to use techniques taught in the class to find vulnerabilities in open source software. The exercise resulted in 44 bugs [Lemos04]. In addition, the NSA's Center of Excellence program provides a unified computer security curriculum to universities that adopt it. While this curriculum is focused on security primarily from a U.S. government perspective, its goal of introducing security topics into all of the major undergraduate Computer Science courses (data structures, compilers, etc.) is a welcome step in the right direction [NSA05].

However, novel education methods will do little to cure the security problem in the short term. Thousands of unaware developers and millions of existing programs must be dealt with. Helping current developers with existing programs is our major motivation for writing this book.

WHY WE WROTE THIS BOOK AND WHY YOU SHOULD READ IT

Our business is helping software vendors ship more secure code. We've been involved in dozens of security testing projects from leading software vendors and the U.S. government. We've also served as consultants to software vendors to help them build more secure code, taught courses on secure coding and security testing, and attended more conferences than either of us can count. While several good books have been written on secure coding, such as *Writing Secure Code* mentioned earlier in the chapter, we found that there were still open questions in the development and testing community. In 2003, we addressed some of these needs with the publication of the book *How to Break Software Security* [Whittaker03], which was the first book on application security testing. The book explored security testing techniques to help software testers find software flaws before their applications shipped. We had an outpouring of support from the software testing community, and that book, along with the accompanying courses and lectures, has helped tens of thousands of software testers to find security vulnerabilities.

Why then another book on software security? The idea for this book came from our own internal security testing needs. Beyond applying the techniques in *How to Break Software Security*, we found that we needed a field guide to security vulnerabilities themselves. We needed this guide to educate both new security testers coming into the company and as a reference for us as we developed our own software. The result was a thorough study of all the reported security vulnerabilities we could find in software. Our search spanned not only vulnerabilities we had exposed through our own testing projects but those in public databases as well, like BugTraq, CVE (the Common Vulnerabilities and Exposures list), CERT (the Computer Emergency Response Team, located at Carnegie-Mellon), and many others. We then started the painstaking process of classifying these vulnerabilities into groups that would be helpful both for finding and fixing vulnerabilities. Some of them clustered naturally, while others had more complicated relationships. At first, the result was an internal pamphlet that we used to help guide our development and testing efforts. Over time, this pamphlet grew into a full-blown guide to security vulnerabilities in software. The material made its way into talks we've given at conferences, and later into our tutorials and magazine articles. Eventually, the demand rose to the level that we felt this information should be disseminated widely in a book. This book is our final product.

TOP FIVE THINGS YOU CAN DO TO IMPROVE YOUR PRODUCT'S SECURITY

1. Validate all user input. The previous sidebar within this chapter and Part III of the book explain why this is important.
2. Assume the *principle of least privilege*. When dealing with a multi-user system or other system with roles and permissions, limit the privilege of an application or routine to just those permissions needed to accomplish its task. Don't, for example, require that your program be run as a superuser because you don't want to worry about file permissions. When you must elevate privilege, limit the privileged application only to the task required.
3. Don't trust the source of your information. Too often, programmers assume that because they wrote both the client and server software, or because they are using a mechanism like SSL to encrypt communication, that they are effectively in control of both sides of a communication. It *is* possible to ensure trust between two applications communicating over the network,

> but it requires a more nuanced solution than just protecting the data in transit or creating an obfuscated protocol.
> 4. Learn from the mistakes of others. Online vulnerability databases like SecurityFocus (*www.securityfocus.com*) or CERT (*www.cert.org*) are a treasure trove of information about bad practices and past mistakes. These sites cover a whole range of software from commercial applications, popular Web sites, and even hardware appliances. System administrators commonly check these sites to see what patches are required to maintain (relative) security within their networks. Developers can benefit from these sites by learning about the mistakes other programmers have made in applications similar to theirs. You might even want to download an exploit or two and try it out on a test system. (Remember, never test against a system that does not belong to you!) By studying an exploit in a hex editor or debugger, you might gain a better understanding of the underlying problem than you can by reading the vulnerability posting, which is often written for a general audience.
> 5. Think about security as a process, not as a goal or an afterthought. During each stage of the development process (whatever that might be for you; we make no claim that one development model is better than another in preventing security vulnerabilities), consider how vulnerabilities can be avoided. For example, while no programming language is free of vulnerabilities, developers should be aware during the language selection process that if C or C++ is chosen, additional safeguards such as the use of safe copy routines are required.

Embedded in these pages are the experiences and knowledge of countless security testers, based on tens of thousands of real software vulnerabilities. While such a guide can never be complete, we focus on the most important security flaws in software and the methods to find, fix, and prevent them. Each chapter deals with one specific class of vulnerability. In addition to detailed explanations and examples, we also provide a summary sheet at the end of each chapter that can be thumbed to and referenced quickly.

The vulnerabilities, techniques, and strategies presented in this book span software in all forms and on multiple operating systems. While most of the examples provided in the chapters are given on the Windows, Linux, and Unix platforms, the underlying vulnerabilities transcend these platforms. The book also addresses vulnerabilities that result from applications that were written in a wide variety of programming languages. While the risk of certain vulnerabilities might be reduced by specific programming languages (such as the unlikely event of a buffer overrun for an application written entirely in Java or C#), most of the vulnerabilities discussed are language independent. Where appropriate, we talk about the differences in programming languages and the risks of a particular vulnerability occurring in applications written in those languages.

Many of the chapters contain source code to illustrate how to find, fix, or prevent a specific vulnerability. For most vulnerabilities the source code examples are written in C, C++, or Java because of the popularity of these languages in the software development community. The exception is Part VI, where HTML, VBScript, JavaScript, PHP, and Perl dominate the examples. The source code for most of the book's examples is included on the CD-ROM that ships with this book.

In many of the chapters we also discuss tools that make the task of finding or fixing vulnerabilities easier. Most of these tools are freely available for download, and where possible we have included them on the CD-ROM that comes with this book.

HOW THIS BOOK IS STRUCTURED

This book was designed to be a reference for developers and testers. Our recommendation is to read it through, cover to cover, once and then keep it handy to reference as you need it. The book is divided into seven major parts. Part I provides background material on software security along with information on security tools that will be used in the book. Parts II through VI are devoted to specific vulnerabilities in software with one chapter devoted to one vulnerability class. Part VII offers some guidance on where we think software security is headed and what types of vulnerabilities are likely to be on the near horizon.

In many of the chapters we have included example source code. Most of this code can be found on the CD-ROM that comes with this book. The CD-ROM also contains free or trial versions of several security tools discussed in the book.

Whenever a chapter points to a resource included on the accompanying CD-ROM the CD icon to the left will appear.

The book also contains note icons.

Note icons like the one on the left highlight common coding mistakes, errors, or other important information.

As previously indicated, Parts II through VI of the book are focused on the software vulnerabilities themselves, with one chapter devoted to each vulnerability. Each chapter takes an in-depth look at the vulnerability and discusses techniques that can be used to find and fix these types of flaws. The general format for each chapter in these sections is as follows:

Describing it: We talk about the vulnerability, give examples of the vulnerability in released software, and talk about the types of software that you are likely to find the vulnerability in.

Finding it: We talk about the tools and testing techniques you can apply to find this problem in software.

Fixing this problem: We discuss techniques that developers can use to fix these flaws in software. In many cases we provide both vulnerable source code and a fixed version of that code. Where appropriate, we also discuss architectural decisions that are likely to prevent this particular class of vulnerability.

At the end of each chapter we also provide a *Summary Sheet*. This is designed to be an easy to refer to distillation of information about the vulnerability. The format for the Summary Sheet is as follows:

Summary Sheet—*Vulnerability*

Problem:

This section contains a single paragraph that summarizes the vulnerability.

Potential Impact:

The possible consequences of this type of vulnerability.

Habitat:

In what types of software and under what conditions this vulnerability is likely to be found.

Tools You Need to Find It:

A list or brief description of the tools that aid in finding this class of vulnerability.

How to Look for It:

A one- or two-paragraph description of the testing techniques that can be used to find this type of vulnerability.

Symptoms of Failure:

This section summarizes the symptoms of this vulnerability in software.

Famous Failures/Exploits:

A summary of one to three famous instances of this vulnerability in released software with references.

The bullet list that follows contains a brief description of each part of this book.

Part I—Introduction: The first part of this book provides the tools you need to fully take advantage of the information presented in the rest of the book. Chapter 2 talks about the fundamentals of software security and the looming threat from attackers. This chapter also discusses the fundamentals of network communications and how data is often manipulated to exploit vulnerabilities in network-enabled applications. Chapter 3 discusses the tools that you need to find and fix many of the security problems in your applications. Many of the tools described in this chapter are used in the discussion of the specific security vulnerabilities in Parts II–VI.

Part II—System-Level Attacks: This part talks about vulnerabilities that occur at the system level and usually stem from poor configuration and architectural decisions made during design and development. Four vulnerabilities are discussed in this part, each with its own chapter. We discuss issues related to dynamic linking and loading, as well as common mistakes in the handling of external resources such as files. We also talk about password and script issues that allow attackers to enter a system and exploit application logic flaws.

Part III—Data Parsing: Arguably one of the most severe software security issues in software is the improper parsing of user data. This part takes an in-depth look at data parsing vulnerabilities and considers the parsing of data that comes directly from the user (through the user interface [UI]) and also data from software's other interfaces, such as the file system and application programming interfaces (APIs). Chapter 8 deals with the notorious buffer overflow. Buffer overflows have traditionally accounted for a significant number of security vulnerabilities [Cowan99], and this chapter discusses how this type of vulnerability can be found, prevented, and fixed. In this part we also discuss format string vulnerabilities and other errors related to the parsing of data.

Part IV—Information Disclosure: This part deals with coding practices that leave sensitive data visible to an attacker. We cover vulnerabilities associated with temporary files, incomplete deletes, exposed data in memory, and visibility through external software components. This part is particularly important for applications that must process sensitive data in any form. Applications that are particularly at risk to the vulnerabilities presented in this part are those that support digital rights management for the protection of documents or other media.

Part V—On the Wire: This part addresses vulnerabilities that can be exploited by either intercepting or tampering with data in transit. Specifically, we look at spoofing vulnerabilities, race conditions, and applications that reveal too much information to an attacker through network-based error messages.

Part VI—Web Sites: This, the longest part in the book, deals exclusively with vulnerabilities that affect Web applications. Web-based applications, be they intranet or Internet facing, must be protected against malicious users. This part presents the most common security issues with Web applications by examining the nine most common vulnerabilities found in applications that run on the Web.

Part VII—Conclusion: The last part (and chapter) of the book takes a look at the emerging trends in software development and where future vulnerabilities are likely to be concentrated. There is a massive push in the industry to try to fix some classes of software vulnerabilities generically at the operating system level. In the Linux world, for example, several flavors of the operating system such as Immunix now have compiled-in stack protection [Cowan99] to guard against stack-based buffer overflows in applications. Similarly, Microsoft (starting in Windows XP SP2) is including stack protection at the operating system level to try to diminish the application buffer overflow threat. Compiler writers are also taking up the challenge by including compiler options to reduce the risk of string-related code flaws resulting in application vulnerabilities. Still others are taking the virtual machine approach to protecting applications such as the Java Virtual Machine (JVM) and Microsoft's push to move applications toward Managed Code. Hardware vendors have even gotten into the game: AMD's 64-bit processor is capable of marking memory pages with a flag that prevents execution of code. While these efforts will hopefully close the door on some types of vulnerabilities, new technologies will undoubtedly open others.

We have chosen not to cover a number of topics, but are worth mentioning. *Physical security*, when discussed in terms of computer security, covers access control by means of locks, badges, cameras, tamper-proofing and tamper-evident systems, etc. While these are without a doubt important mechanisms for protecting information, we are of the view that if an attacker gains unmonitored, unrestricted physical access to your computer, it's not your computer anymore. As a result, there aren't a lot of things a developer can do to prevent or protect against these kinds of attacks. If you're interested in this aspect of security, Ross Anderson's *Security Engineering* [Anderson01] covers these issues in detail and is well worth reading. (We especially enjoyed Chapter 11, "Nuclear Command and Control," which is a fascinating discussion of security when failure would have unthinkable consequences.)

Social engineering is the term computer security professionals use to describe circumventing security through psychological manipulation of a user, or through outright trickery. Mass-mailing the customers of a bank and asking them to re-enter their account information into a bogus Web site is a good example of social engineering. (It's actually a variation of social engineering called *phishing*, which is discussed in Chapter 22.) So is lying to an administrator in an attempt to have a password reset. Some safeguards can be applied to prevent social engineering (such as the "mother's maiden name" technique used by some banks), but no clear-cut technological solution exists.

We also don't discuss design of secure networks. While 90 percent of all vulnerabilities over the next several years will be related to flaws in software (according to Joe Pescatore of Gartner Group, [Pescatore03]), most of the books about computer security are oriented towards securing networks. We feel that while knowledge of secure networks is useful, it is much more important that developers be aware of the underlying causes of security vulnerabilities: bugs in software.

Finally, we have not made this a how-to manual for writing securing software. It is not a code cookbook, though code examples are found throughout our chapters. We have made this decision for two reasons. First, examples are dangerous in the hands of an unknowledgeable developer. They tend to create a false sense of security, because it is assumed that an example is safer than writing your own code. You might have nuances to your system that we cannot anticipate; they might render the protections demonstrated in the example useless, or might conceal a more serious underlying vulnerability. Second, we know many developers who are just now beginning to think about security, but might already have a significant amount of legacy code developed. In this circumstance, understanding and identifying existing vulnerabilities is much more important than having implementation examples. If the reader does need secure programming examples that are reasonably free of errors, we recommend several books. The aforementioned Michael Howard and David Leblanc's *Writing Secure Code* [Howard02] (now in its second edition) and John Viega and Gary McGraw's *Building Secure Software* [McGraw01] contain a wealth of source examples a developer can use to implement pieces of a secure application. Neither of these books contains many examples of Web applications, however. Web developers should consider Sverre Huseby's *Innocent Code* [Huseby04].

WHO WE ARE

Herbert Thompson is Director of Security Technology at Security Innovation (*www.securityinnovation.com*) and also serves on the Graduate Faculty of the Florida Institute of Technology. Herbert is co-author (with James Whittaker) of *How to Break Software Security: Effective Techniques for Security Testing* (Addison-Wesley 2003) [Whittaker03] and is the author of numerous papers on software security and testing. At Security Innovation, he works on developing and teaching security testing techniques, and in this role he has taught security testing to developers at over 50 companies and is frequently booked to give courses to the U.S. military.

Scott Chase is Security Architect at SI Government Solutions (*www.sigovs.com*) and manages key research projects for the U.S. government. As a result, he spends much of his time in the laboratory, developing and using the techniques described in this book to uncover vulnerabilities in a variety of critical software systems. He has also worked as a university researcher in information security and as a software tester in industry.

REFERENCES

[Anderson01] Anderson, Ross. *Security Engineering: A Guide to Building Dependable Distributed Systems.* Wiley, 2001.

[Cert04] "CERT/CC Statistics 1998–2004." CERT Coordination Center Web site at Carnegie-Mellon University. Online at *www.cert.org/stats/cert_stats.html*.

[Cowan99] Cowan, Crispin, et al., "StackGuard: Automatic Adaptive Detection and Prevention of Buffer-Overflow Attacks." *Proceedings of the 7th USENIX Security Symposium.* San Antonio, Texas, January 26–29, 1998.

[Howard02] Howard, Michael and LeBlanc, David, *Writing Secure Code,* 2nd Ed. Microsoft Press, 2002.

[Huseby04] Huseby, Sverre. *Innocent Code: A Security Wake-up Call for Web Developers.* Wiley, 2004.

[Lemos04] Lemos, Robert. "Students uncover dozens of UNIX software flaws." *CNET News* 12/15/2004. Online at *http://news.zdnet.com/2100-1009_22-5492969.html*.

[Lipner04] Lipner, Steve, "The Trustworthy Computing Security Development Lifecycle." Proceedings of the 20th Annual Computer Security Applications Conference (ACSAC '04), Tucson, Arizona, December 6–10, 2004.

[McGraw01] McGraw, Gary, and Viega, John. *Building Secure Software.* Addison-Wesley. Summer, 2001.

[NSA05] National Security Agency Centers of Academic Excellence. Online at *www.nsa.gov/ia/academia/caeiae.cfm*.

[Pescatore03] Pescatore, Joe. Gartner. "CIO Alert—Follow Gartner's Guidelines for Updating Security on Internet Servers, Reduce Risks." February 2003.

[Whittaker03] Whittaker, James and Thompson, Herbert. *How to Break Software Security: Effective Techniques for Security Testing.* Addison-Wesley Publishing Co., 2003.

2 Security Background

In This Chapter

- Hacker versus Cracker versus Attacker: The Language of Computer Security
- Legal and Ethical Issues Surrounding Computer Security
- Networking Basics
- Networking References
- References

To the average programmer, there's something mysterious about security. Perhaps it's because, unlike most other aspects of the software industry, computer security is a shadowy, lawless discipline. Movies, TV, books, and the Internet are captivated by the romance of the "hacker," a uniquely twenty-first century villain, with near absolute powers in a realm that is foreign and frustrating to the average person. In the television world, legions of these loosely associated miscreants wage war every day against the government, banks and financial institutions, giant corporations, and authority in general. Hackers, in this world, are always loosely associated; that kind of antisocial, loner type who could never form close associations with anyone.

The make-believe world of cyber-wars and hacker heroes does not entirely overlap with reality. Several reasons exist for this. First, like police work or soldiering, hacking is a lot more glamorous in books and TV than in real life. Passwords cannot often be guessed in three tries. (See Chapter 5.) Though there have been several attempts to create "Hacker GUIs" (the most famous being the BackOrifice interface for Windows NT), these have been poor copies of the art they are trying to imitate. Second, though most hackers feel that they are making a political statement against the establishment by their actions, they don't frequently succeed against their main adversaries; folks at the Department of Defense or CIA have serious systems and information to protect, and don't leave it exposed to the casual hacker. In fact, the *WarGames* notion that nuclear missiles can be launched by computer seems to conflict with the *Crimson Tide* notion that they must be launched with mechanical keys. The result is that the most frequent victims of hackers are the "weak" of cyberspace: universities and libraries committed to academic freedom, small businesses that cannot afford to upgrade to the latest software, and unsuspecting end users. The recent rise in "phishing" attacks, in which an attacker attempts to trick a victim into giving up a credit card number or other information, is a testimony to this. It is also disturbingly reminiscent of telephone scams directed at the elderly. Hardly the stuff of television.

Stripped of its Hollywood glamour, hacking makes use of the same skills, knowledge, and technology as any other kind of programming. For this reason, we feel that the ordinary programmer can learn how to prevent exploitation of his own software without much difficulty. This chapter is meant to be an introduction to some of the concepts of computer security that will be important to understanding later chapters. Grasping these basics, along with an understanding of the individual vulnerabilities described in the rest of the book, will enable you to compete with the wily attacker (TV or otherwise) and use the advantage you have by virtue of the superior understanding you have of your own application. One final note: some of our readers who are not programmers might feel that they are in over their heads. Don't worry; many of the techniques outlined in this book do not require advanced programming skills. If you feel you need to do some catching up to understand the rest of this book, the references section at the end of the chapter contains some good places to start.

HACKER VERSUS CRACKER VERSUS ATTACKER: THE LANGUAGE OF COMPUTER SECURITY

Like any technical field, security has its own jargon. It borrows heavily from the rest of the computer science world, as well as the physical security (safe and lock) domain and some military vernacular. This section is our mini-glossary of terms you need to understand to follow the rest of this book.

We use *attacker* to mean someone who attempts to bypass the security of a piece of software to gain advantage over it. This person might be a *hacker* or *cracker*, both of which have different meanings from the one we intend. A hacker is any person who "hacks"; that is to say, writes computer programs, especially in an undisciplined, disorganized way. Most programmers pride themselves on being hackers in this sense and take offense to the lovable pejorative for our profession being applied to criminals and ne'er-do-wells. Most would prefer that these folks be referred to as crackers, because they "crack" through the security surrounding applications. However, we avoid this term because we tend to think of only certain types of attackers as crackers. Cracking a program usually means bypassing the licensing or copy protection in a program, one kind of application security. It can also mean discovering the encryption key or password associated with a piece of data, through brute force means, akin to safe cracking. Meanwhile, a *script kiddie* is a person who fancies himself a hacker, but prefers to download other people's exploits and use them. Script kiddies talk in *leetspeek*, which is a way of obfuscating words using numbers as letters. "H4x0r" is pronounced "hacker" in leetspeek. Leetspeakers have their own search engine; check out *www.google.com/intl/xx-hacker/*.

A *vulnerability* is a bug in software that enables an attacker to bypass security. Programmers, not attackers, are responsible for vulnerabilities, usually due to poor security knowledge. On the other hand, an *attack* is a technique attackers have developed to identify whether a vulnerability exists in a particular piece of software. An *exploit* is a method or piece of code that takes advantage of a vulnerability to accomplish an attacker's goal. Sound confusing? Often, the terms are used interchangeably, especially in situations where only one attack exists for a given vulnerability, and one exploitation of that attack. We have tried to keep the distinctions as clear as possible, even when some vulnerabilities are more commonly referred to as attacks (e.g., cross-site scripting).

Spoofing means impersonating a user, a machine, or some other entity as a means of tricking the security logic in an application. Much of security depends on trust, and often an application trusts that an entity is who (or what) it says it is, or that the entity cannot change in the middle of a session. *Hijacking* means an attacker takes control of one side of a two-sided transaction in the middle of that transaction. Hijacking is frequently accomplished by means of a *man-in-the-middle attack*, in which the attacker manages to gain control of a location (such as a function or a network node) that lies between the two parties in the transaction.

Information disclosure means an attacker is able to see information belonging to another user or that he shouldn't otherwise see. One of the main principles of security is protecting the privacy of information; information disclosure would violate this principle. *Escalation of privilege* refers to an attacker gaining a higher level of access to information or functionality than he is authorized; for example, from an ordinary user to an administrator. *Denial of service*, on the other hand, means preventing other users from accessing a piece of information or a resource, without necessarily gaining any privilege. Causing service applications to crash irrecoverably is a frequent means of accomplishing a denial-of-service attack.

Cracking, as we mentioned previously, means circumventing the copy or license protection associated with a piece of software. "Cracks," programs that allow the software to be installed or copied in an unlicensed manner, are often posted alongside exploits on underground security Web sites. *Reverse engineering* means partially recovering the design or algorithms of a program to make modifications to it. Reverse engineering techniques are often required both for cracking and development of exploits. Cracking can also mean recovering a password or encryption key that protects an application.

A *virus* is a piece of software that is capable of causing malicious damage while at the same time spreading itself. Viruses work by attaching themselves to a file, usually an executable. Viruses typically spread when the infected file is shared. A *worm*, on the other hand, spreads by sending copies of itself to other users or machines. Many folks use these terms interchangeably. A *Trojan horse* is made to look like a benign program, but actually implements malicious functionality. Attackers will often "Trojan" applications like the system login program to obtain usernames and passwords. If the Trojan permits an attacker to regain access to the system at a later time, it is called a *back door*.

Some attacks rely on human psychology rather than technology to accomplish their purpose. *Social engineering* refers to the process of tricking a user into volunteering information that an attacker can use to his advantage. Bribery and fictitious login screens are both examples of social engineering. When social engineering is employed on a large scale to obtain passwords, credit cards or other personally identifiable information, it is called *phishing*. Notice that phishing is essentially the same word as fishing, with a more exotic spelling. This is because a "phisherman" casts out many lines in the form of a spam e-mail or instant message, in the hopes of catching a few of the gullible "phish."

There's some constructive terminology within security as well. A *firewall* is an application or appliance that limits the flow of information into or out of a system to diminish the risk of a successful attack. An *intrusion detection system*, like a monitored burglar alarm, actively monitors a system for signs that an attack is under way. *Penetration testing* or *red teaming* refers to the act of simulating an attack to design safeguards. An *ethical hacker* often carries out penetration tests. We do not consider ourselves ethical hackers. Ethical considerations aside, we call ourselves *security testers* because we perform the function that a tester would in any software development process, though we specifically focus on security. If you have a software company, you should hire programmers or testers with security knowledge, or train your existing folks, rather than hire hackers who have reformed (or are looking to reform) their ways.

LEGAL AND ETHICAL ISSUES SURROUNDING COMPUTER SECURITY

To certain narrow-minded folks, this is a book about hacking. Without a doubt, some of the techniques taught in this book can be used to gain unauthorized access to another person's computer system. Our position is that such action is always ethically wrong, and can be illegal in some circumstances. This section summarizes some of the specific legal issues surrounding security testing. But first, we owe an explanation as to why we feel it is appropriate to include such techniques in a book written for developers. The fact is that all of the best attackers out there are familiar with some or all of these techniques, and many of the not-so-good ones are as well. The "black hat" community (the attackers who are interested in engaging in illegal or malicious activity) has a wealth of information about attack techniques available through Web sites, newsgroups, and online magazines such as *Phrack* (*www.phrack.org*) and *2600* (*www.2600.com*). Developers who write the software that is being exploited by these groups have fewer resources available in this respect. We spend a good part of our time educating developers through online and onsite training, presentations at trade shows and conferences, and in private consulting. However, we feel that books are necessary to reach an audience broad enough to bring about real change. It is a risk that enterprising attackers will take advantage of our techniques to compromise unsuspecting victims. However, the upside that these techniques will cease being effective because of increasing developer awareness and testing is worth that risk.

Federal Laws Related to Illegal Computer Use

Using a vulnerability to gain unauthorized access to a computer system is illegal under Title 18, Sec. 1030 of the United States code. Other countries and many states have similar laws. Sec. 1030 [USC05a] defines unauthorized access as:

> (5)(A)(i) knowingly causes the transmission of a program, information, code, or command, and as a result of such conduct, intentionally causes damage without authorization, to a protected computer;
>
> (ii) intentionally accesses a protected computer without authorization, and as a result of such conduct, recklessly causes damage; or

(iii) intentionally accesses a protected computer without authorization, and as a result of such conduct, causes damage;

Damage is defined as:

(i) loss to 1 or more persons during any 1-year period (and, for purposes of an investigation, prosecution, or other proceeding brought by the United States only, loss resulting from a related course of conduct affecting 1 or more other protected computers) aggregating at least $5,000 in value;

(ii) the modification or impairment, or potential modification or impairment, of the medical examination, diagnosis, treatment, or care of 1 or more individuals;

(iii) physical injury to any person;

(iv) a threat to public health or safety; or

(v) damage affecting a computer system used by or for a government entity in furtherance of the administration of justice, national defense, or national security.

Additional portions of Sec. 1030 make provisions for theft of government and defense information and information relating to banking activity such as account numbers, and other sections of the code deal with illegal computer activity directed at communication systems such as the national telephone network or Internet. This statute covers most kinds of illegal computer activity, and it is reasonable to suspect that most unauthorized access could be prosecuted under this law.

The Economic Espionage Act, Title 18, Sec. 1831–9, makes it illegal to sell a company's proprietary information. Intellectual property, such as confidential documents, plans, and source code, falls under this provision. So does personal information about a business' customers [USC05b].

The Digital Millennium Copyright Act [DMCA98] makes it illegal to reverse engineer software for the purposes of circumventing copy protection. Many people erroneously believe that this prohibits all reverse engineering, or at any rate, reverse engineering to circumvent security. Because the DMCA (as it is often called) is intended to protect intellectual property rights owners from illegal copying, many kinds of reverse engineering (such as creating a video game cheat) are not covered by DMCA. In fact, reverse engineering for security testing is expressly permitted by the DMCA:

(j) Security Testing.—

(1) Definition.— For purposes of this subsection, the term "security testing" means accessing a computer, computer system, or computer network, solely for the purpose of good faith testing, investigating, or correcting a security flaw

or vulnerability, with the authorization of the owner or operator of such computer, computer system, or computer network.

(2) Permissible acts of security testing.— Notwithstanding the provisions of subsection (a)(1)(A), it is not a violation of that subsection for a person to engage in an act of security testing, if such act does not constitute infringement under this title or a violation of applicable law other than this section, including section 1030 of title 18 and those provisions of title 18 amended by the Computer Fraud and Abuse Act of 1986.

(3) Factors in determining exemption.— In determining whether a person qualifies for the exemption under paragraph (2), the factors to be considered shall include—

(A) whether the information derived from the security testing was used solely to promote the security of the owner or operator of such computer, computer system, or computer network, or shared directly with the developer of such computer, computer system, or computer network; and

(B) whether the information derived from the security testing was used or maintained in a manner that does not facilitate infringement under this title or a violation of applicable law other than this section, including a violation of privacy or breach of security.

(4) Use of technological means for security testing.— Notwithstanding the provisions of subsection (a)(2), it is not a violation of that subsection for a person to develop, produce, distribute, or employ technological means for the sole purpose of performing the acts of security testing described in subsection (2), provided such technological means does not otherwise violate section (a)(2).

Some software license agreements prohibit reverse engineering. However, the enforceability of these agreements is limited by two factors. First, because copyright law is constitutional law, it is not clear to what extent a vendor can limit constitutionally protected fair use, especially if no malicious intent exists on the part of the reverser. Additionally, in the case of "clickwrap" licenses (where a user simply presses "yes" to continue or something similar) it is questionable whether a binding contract exists between the vendor and the consumer. This is especially true of mass-market software when the vendor might not be aware of the identity of the consumer. Cem Kaner and David Pels discuss this topic in great detail in their book *Bad Software* [Kaner98].

We are not attorneys and do not know the laws in every jurisdiction in the world. The laws in your jurisdiction might be more or less strict. Our advice is that if you suspect what you want to do is illegal, you probably shouldn't do it.

Ethical Reporting of Security Vulnerabilities

To our knowledge, U.S. law does not require individuals to report vulnerabilities that are discovered to the vendor of a software application, or to the owner of an affected system. This might be different in other jurisdictions. The law notwithstanding, it is good practice to do this in as discreet and timely manner as possible. Most reputable security researchers have adopted an ethical disclosure policy, in which vulnerabilities are reported first to the vendor whose software contains the vulnerability and to the rest of the community only when that vulnerability is patched.

We do not approve of "full disclosure." This practice, observed by some black hats and pseudo-black hats (sometimes called "gray hats," these individuals engage in both legal and illegal computer security research), entails reporting a vulnerability to the whole software community before a vendor has the opportunity to release a fix. Full disclosure proponents believe that consumers of software have a right to know about vulnerabilities in their own systems as soon as they are discovered. Additionally, full disclosure forces recalcitrant vendors to fix vulnerabilities in a timely fashion. However, this kind of disclosure invariably leads to unsuspecting users being compromised. Because the cost of viruses and worms facilitated by vulnerabilities can be staggering, we feel full disclosure is irresponsible.

Akin to full disclosure is the "ticking time bomb" approach. In this situation, a researcher keeps a vulnerability private for a set amount of time and then releases it to the public regardless of whether a patch has been issued. While this is intended to force the vendor to patch quickly, it is essentially delayed full disclosure.

One of the worst practices is that of researchers who find a vulnerability and then ask for money, either in the form of a testing contract or an outright bribe, to keep a vulnerability quiet. This ugly practice does happen from time to time in our industry. We view this as extortion and, insofar as the vulnerability was found specifically to blackmail the vendor, it might constitute racketeering as well. You should never pay for security testing from a company that engages in these practices. How do you know that they will keep your secrets even after they are paid?

On the other hand, we think that the practice of some vendors of silently patching vulnerabilities, without providing subsequent notification to the user, is also irresponsible. By failing to disclose that a vulnerability exists, the vendor is effectively hiding the real security of their product from their users.

NETWORKING BASICS

To understand quite a bit of this book, a working knowledge of TCP/IP networking is needed. Skip this section if you feel you have an intimate knowledge of how

TCP/IP and protocols such as HTTP operate. The information presented here is by no means complete; volumes can be written about just the basic architecture. (The late W. Richard Stevens did write five volumes on the subject under the titles, *TCP/IP Illustrated, Vol. 1–3* and *Unix Network Programming, Vol. 1 and 2* [Stevens93], [Stevens98], [Stevens03]. These are must-haves in the security and networking world.)

The underlying premise of the TCP/IP family of protocols is its multilayer model. Though the OSI officially recognizes seven layers, in practice there are only five, and we worry about only four of them. Each layer performs a particular function, and is independent of both the layer above and below it. This means protocols can run over a wide variety of hardware and software configurations under different conditions, without apparent differences to the programmer or end user. Each higher level protocol is *encapsulated* within the lower protocol—that is to say, the "data" portion of the lower protocol contains the whole packet from the higher protocol, much like the *matryoshka* nesting dolls from Russia. Figure 2.1 shows a typical TCP/IP encapsulation.

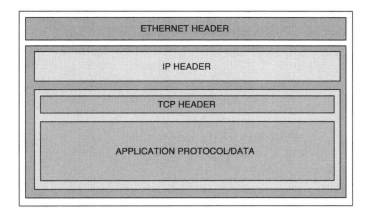

FIGURE 2.1 Typical TCP/IP encapsulation.

SOME (PHYSICAL) THINGS YOU MIGHT FIND ON A NETWORK

A *hub* is a device that repeats physical layer packets among multiple devices. Most computers are connected to other computers by means of a hub. When a packet is transmitted by the Ethernet (or other protocol) card on a host to the hub, it is rebroadcast to all the other nodes on the hub. The other hosts generally ignore traffic that is not addressed to them.

A *switch* is similar to a hub, except that it has software that knows the hardware (MAC) address of each machine connected to it. As a result, it is able to transmit a packet only to the host for which it is destined. This reduces noise on the network and also improves security because an attacker does not by default receive all of the network's traffic through his machine. (Some techniques taught in Chapter 16 make this an imperfect solution, though.)

A *router* routes traffic between two networks. Host systems are configured with a default route—an IP address to which packets destined for to another network or to the Internet are sent. These hosts send traffic destined for the remote network to the router using the remote system's IP address but the router's physical address. It then passes the traffic to the other side. A router might also use *masquerading*, sometimes called *Network Address Translation*. In this configuration, individual systems on the network have private IP addresses, which are not routable via the Internet. The router then disguises the traffic from all of these systems as though it came from a single host, the router. This preserves privacy and security, because host systems are not directly addressable via the Internet, and also preserves IP addresses.

A *firewall* is a filter for network traffic. Firewalls are generally configured to block protocols or individual ports, but some can do more sophisticated filtering based on the contents of the traffic. For example, we might permit incoming traffic to our network only via port 80. In this case, a user would not be able to access systems within our network via FTP, SMTP, or any other protocol. The best practice is to limit traffic to as few ports as possible, enabling only those that are needed for specific purposes.

Intrusion detection systems are systems that attempt to alert an administrator to a possible attack on the network. Because this is not an easy thing to detect, these systems tend to have many false positives (cases where no attack occurs but the system reports an attack anyway) and false negatives (cases where an attack occurs but the system is unable to recognize it).

The *physical layer* (that's the one we don't concern ourselves with) refers to the way in which communication protocols are electrically and logically implemented in hardware. Manufacturers of networking hardware including routers, hubs, switches, and Ethernet cards have to worry with these Layer 1 issues, which include voltage levels, physical timeouts, maximum distances, etc.

The *data link layer* refers to the software that frames data to be transmitted over the physical medium. These mediums include Ethernet, ATM, FDDI, and Token Ring; the device drivers for these pieces of hardware implement the data link layer.

This is the lowest level a programmer would concern herself with. Ethernet is by far the most common data link layer protocol in an end-user environment. In an Ethernet, each device connected to the network has a unique *Media Access Control address*, commonly called a MAC address. This 6-byte hexadecimal number is supposed to be different on every machine because each manufacturer is assigned prefix codes by the Internet Assigned Numbers Authority (IANA), the same agency that also regulates IP addresses and port numbers. Each manufacturer might make cards beginning only with assigned range of prefix codes. However, some manufacturers occasionally recycle MACs, while others allow the MAC to be changed in software. As a result, you can't *count* on MACs being unique for security purposes.

The format of an Ethernet packet is very basic and is shown in Figure 2.2. Its fields are:

- The 6-byte destination MAC address
- The 6-byte source MAC address
- A 2-byte type code
- The data, which can be between 46 and 1500 bytes long
- A 4-byte checksum

Destination MAC (6 byte)	Source MAC (6 byte)	Type (2 byte)	Data (46-1500 bytes)	Checksum (4 byte)

FIGURE 2.2 Format of an Ethernet frame.

If the actual data size is less than 46 bytes, the data area is padded with zeroes. The type code is used to describe the protocol encapsulated within the data field. The most common values of this field are 0×0806, the code for an Address Resolution Protocol (ARP) request, and 0×0800, the code for IP.

Address Resolution Protocol is a Layer 3, or *network layer*, protocol that runs over Ethernet. This protocol is used to map MAC addresses to IP addresses within the local network. Each time a machine wants to send an IP packet to another machine, it must first learn the MAC address in order to correctly compose the Ethernet frame. It does this by broadcasting an ARP packet to every machine on the network. Only the machine with the corresponding correct IP address should reply to the ARP request. If a packet is destined for a machine outside the local LAN, an ARP request would be sent instead to the address of the local gateway, which must be specified in advance. *ARP cache poisoning* occurs when a malicious ARP daemon on the local network responds to every request as though it were its own. By spoofing ARP replies in this manner, an attacker can use his machine as a "man in the middle" between the sender and receiver.

The Internet Protocol (IP) is the main networking layer protocol. It functions like the postman, delivering packets to different machines based on their IP address. On the "real" Internet, IP addresses are also unique thanks to the IANA. Often, private IP addresses are used along with *masquerading*, which translates these private IPs to real addresses and vice versa. An IP packet is quite a bit more complicated than an Ethernet packet, because it is intended to work in a much wider variety of circumstances. Figure 2.3 shows the format of an IP packet. The fields of an IP packet are:

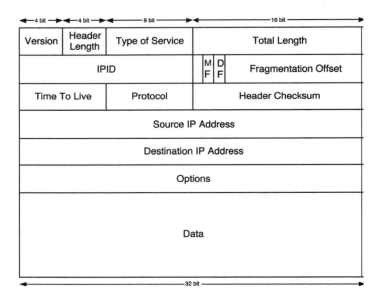

FIGURE 2.3 Format of an IP packet.

- The 4-bit protocol version. This is always "4" in IPv4, the most common kind of IP that is used by the Internet. In IPv6, this value is, you guessed it, "6."
- The 4-bit header length. This also never varies in IPv4.
- The type of service byte. This can have one of several values; routers sometimes prioritize traffic based on the type of service.
- The total length of the packet in 2 bytes.

- The 2-byte IPID, which pseudo-uniquely identifies this packet to the receiver. This eliminates duplication and out of order problems.
- The flags and fragmentation offset, 2 bytes. These are used if the data is too large to be contained in one packet.
- The 1-byte Time To Live. This is decremented by each router a packet passes through to prevent infinite loops.
- The 1-byte protocol number. This is 6 for TCP and 17 (11 hex) for UDP.
- The header checksum, 2 bytes.
- The 4-byte source IP address.
- The 4-byte destination IP address.
- Miscellaneous options.
- The data.

Machines are able to transmit a packet to its final destination by means of routing rules. These rules, contained within each gateway, describe which machine to forward a packet on to in order to try to reach the destination host. On an end-user machine, typically you have four rules. First, packets destined for the loopback interface (127.0.0.1) and the local IP address are not routed; the networking stack within the machine handles these. Packets destined for machines on the local network are routed directly to them by resolving their MAC addresses using ARP. Finally, all other packets are routed to the default gateway, which is a machine on the local network. This machine presumably knows the address of another gateway downstream, as well as any peer local area networks to ours.

The process for determining what machines are on the local network and what machines are beyond the gateway is called *subnetting*. A subnet mask specifies how many bits of the IP address are held in common by all the machines on a local network. A subnet mask of 255.255.255.0 means all machines on the local network will have the same first 3 bytes as our machine. Subnets can be as small as one machine, for point-to-point networks, or as large as 16 million machines for a class A network. In practice, subnets larger than 255.255.0.0 are not used outside the main Internet backbones.

This organization of machines has proved very practical for the Internet, but has some limitations. For example, you cannot know with certainty whether a machine exists when a packet is sent to it. To compensate for this, the final gateway that cannot route the packet further returns an error message to the sender. This ICMP_UNREACHABLE message is akin to the "no such addressee" notice you would write to the mailman if you received someone else's mail. Additionally, you have no way to certify the authenticity of a packet when it reaches its destination. This, combined with the use of rerouting gateways, opens the possibility for a variety of spoofing and man-in-the-middle attacks. Additionally, IP addresses are hard to remember and might change. For this reason, the Domain Name Server (DNS) protocol is used to map logical names (like *www.whitehouse.gov*) to their corresponding IP addresses.

As the network layer protocol, IP can transmit packets only between nodes on the network. A network node is typically represented by a single machine. The next layer up, the *transport layer*, manages the reliability of connections, as well as ensures that the packet is received by the application it is destined for. The Transmission Control Protocol, TCP, is the main transport layer protocol in the Internet scheme. TCP is capable of synchronizing communication between the two applications to ensure that data is received in the correct order with no missing information. It does this by means of a *sequence number* and *acknowledge number*. With each transmission, an application increases its own sequence number, and responds with 1 + the remote host's acknowledge number. If any packets are transmitted out of sync, the sequence and acknowledge numbers will not match. TCP also uses a system of state flags to initially synchronize the two applications. When an application first tries to connect to another, it sends a packet with an initial sequence number and the synchronize (SYN) flag set. The remote application acknowledges this with both the SYN and acknowledge (ACK) flag set, along with its own initial sequence number and acknowledgement. The original sender then sends a packet with the ACK flag only set, along with an acknowledgement of the remote initial sequence number. This is called the *three-way handshake*.

The fields of the TCP packet are set up to accommodate this process. They are:

- The 2-byte source port number
- The 2-byte destination number
- The 4-byte sequence number
- The 4-byte acknowledge number
- The 4-bit header, six 1-bit flags and a reserved area
- The 2-byte window size
- The 2-byte checksum
- The 2-byte urgent pointer
- Miscellaneous options
- The data

Figure 2.4 shows the format of a TCP packet.

A TCP packet is routed to the correct application because only one application at a time can use a port number on a particular machine. The lower value port numbers are used for common services and are assigned to particular applications. Historically, ports with numbers less than 1024 were considered "trusted." A connection originating from one of these port numbers was considered to be from a previously authenticated user, because non-root users cannot open sockets on these

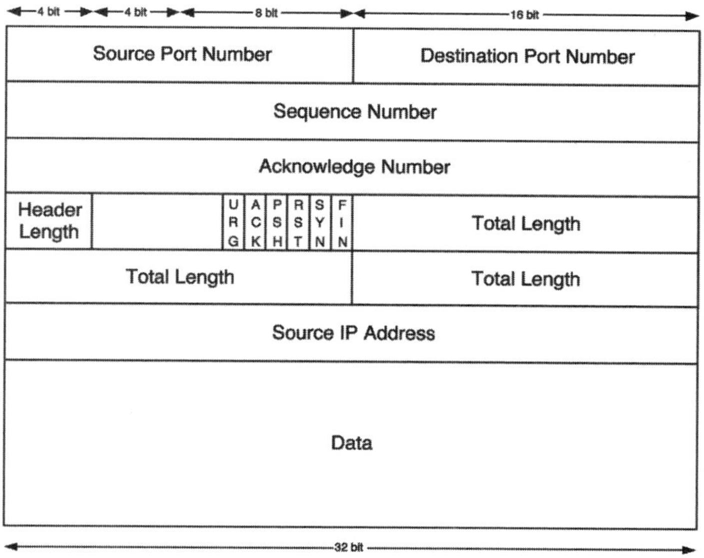

FIGURE 2.4 Format of a TCP packet.

ports. This technique was used by `rlogin` among other things. The flaw in this approach is that it is possible for an attacker to spoof the identity of a remote system, as described in Chapter 16. For example, Web servers, which use the HTTP protocol, are typically assigned port 80. Mail servers that use the Simple Mail Transport Protocol (SMTP) protocol listen on port 25. The higher port numbers are dynamically assigned to client applications that do not need a dedicated port number.

A simpler network layer protocol, the Universal Datagram Protocol (UDP) is also frequently used in networking. However, because it lacks the sophisticated synchronization capability of TCP, it is more frequently used for local, system-level protocols like remote procedure call (RPC) and DNS. (It is assumed that the application, not the protocol implementation, takes care of retransmission of lost UDP packets.) A UDP packet consists only of five fields, as shown in Figure 2.5:

- The 2-byte source port number
- The 2-byte destination port number
- The 2-byte packet length
- The 2-byte checksum
- The data

The individual *application layer* protocols are transmitted over TCP/IP. Some of these protocols are human-readable text, while others are binary. Many are

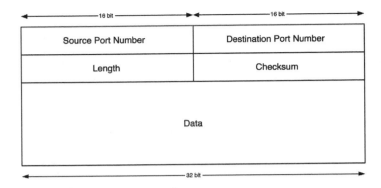

FIGURE 2.5 UDP packet format.

documented through international standards or the Internet Engineering Task Force's Request for Comment (RFC) documents (online at *www.ietf.org/rfc.html*). However, some protocols are undocumented and proprietary, or have proprietary extensions. Table 2.1 lists some of the more common TCP application protocols and their port numbers.

TABLE 2.1 Common TCP Application Port Numbers

Application Protocol	Port Number
File Transfer Protocol (FTP) non-passive transfers	20
File Transfer Protocol (FTP) control stream	21
Secure Shell Protocol (SSH)	22
Telnet	23
Simple Mail Transfer Protocol (SMTP)	25
Domain Name Server Protocol (DNS)	53
HyperText Transfer Protocol (HTTP)	80
Post Office Protocol v.3 (POP3)	110
NetBIOS (Windows Networking) Name Service	137
NetBIOS (Windows Networking) Datagram	138
NetBIOS (Windows Networking) Session Protocol	139
Internet Message Access Protocol (IMAP)	143
Secure HTTP over Secure Sockets Layer (HTTPS/SSL)	443

HTTP, used by Web servers and Web browsers, is among the most simple. An *HTTP request*, destined for a Web server, is a simple text command such as

```
GET /index.html HTTP/1.0
```

followed by a carriage return and line feed character. The preceding command would be used to fetch the file `index.html` from the root directory of the Web server. Alternatively, we could have used GET / HTTP/1.0 and allowed the Web server to select the default page to return. The Web server replies with the requested file within the same connection. It first transmits the *MIME type* of the file, as well as the length. The MIME type instructs the browser on how to handle the file. In a typical HTTP response, this would look like

```
Content-type: text/html
Content-length: 308

<HTML><HEAD><TITLE>My web page</TITLE>
. . .
```

where the file was 308 bytes in length (this varies from file to file). Notice the actual Web page follows the Content-length directive. Additional directives might appear in the response. However, the body of the message is always separated from them by a single blank line at the end.

Many additional protocols run over IP. Two of the more common are the Internet Control Message Protocol (ICMP), which is used to communicate error, control, and information messages (such as non-existent host errors) between machines, and the Internet Group Management Protocol (IGMP), which is used by hosts and routers to manage their multicast groups in situations where multicasting is required.

NETWORKING REFERENCES

As we mentioned previously, the protocols that make up the TCP/IP family are far too many to document here. We recommend that you get a separate book on networking. *TCP/IP Illustrated* by W. Richard Stevens [Stevens93] is considered one of the best books for low-level TCP/IP. If you want a gentler book, Uyless Black's *TCP/IP and Related Protocols* [Black91] is a good introduction.

REFERENCES

[Black91] Black, Uyless. *TCP/IP and Related Protocols.* McGraw-Hill Education, 1991.

[DMCA98] "The Digital Millennium Copyright Act: U.S. Copyright Office Summary." U.S. Copyright Office, Library of Congress. Online at *www.copyright.gov/legislation/dmca.pdf*, December, 1998.

[Kaner98] Kaner, Cem and Pels, David. *Bad Software: What to Do When Software Fails.* Wiley, 1998.

[Stevens93] Stevens, W. Richard (Vols. 1–3) and Wright, Gary (Vol. 2). *TCP/IP Illustrated.* 3 Vols. Addison-Wesley, 1993 (Vol. 1), 1995 (Vol. 2), and 1996 (Vol. 3).

[Stevens98] Stevens, W. Richard. *Unix Network Programming, Vol. 2,* 2nd Ed. Prentice Hall PTR, 1998.

[Stevens03] Stevens, W. Richard, et al. *Unix Network Programming, Vol. 1,* 3rd Ed. Addison-Wesley, 2003.

[USC05a] "Economic espionage." *Title 18 U.S. Code, Sec. 1831,* 2005 Ed. Online at *www.gpoaccess.gov/uscode/*.

[USC05b] "Fraud and related activity in connection with computers." Title 18 U.S. Code, Sec. 1030, 2005 Ed. Online at *www.gpoaccess.gov/uscode/*.

3 Some Useful Tools

In This Chapter

- Security Scanners
- Hacking and Cracking Tools
- Reverse Engineering Tools
- Commercial Tools
- For More Information

Security testing might seem like black magic. Many people assume that hackers, white hat and black hat, possess some kind of mystical knowledge about the internals of applications and systems that reveals security vulnerabilities to them. Film and newspapers, and to a large extent the hackers themselves, seek to perpetuate this image to preserve a notion of elitism and sophistication that sets them apart from ordinary programmers and users. In fact, the hacker has an enormous number of tools at his disposal that make finding and exploiting vulnerabilities easy. Most of the tools are freely available from mainstream sources; it is no longer necessary to have access to an underground bulletin board or Web site in order to obtain them. Additionally, most programmers possess all of the knowledge and skills to use these tools as effectively (if not more effectively) than hackers. The reality is that the most destructive of "black hat" hackers are usually less technologically savvy than the average professional software engineer.

This chapter describes some of the tools a hacker uses in finding vulnerabilities. If you scan the section headings, you'll find that many of these tools are already familiar to you. Where possible, we have selected open source or freeware tools as examples in each category. The hacker doesn't break the bank for software tools; neither should you. At the end of the chapter, we describe some of the better commercial tools on the market. Use these if you require support, if your company prefers closed source or proprietary software, or if you're looking to make an investment in security testing tools.

Most of the tools we've discussed are available online, and evaluation versions are available for most of the proprietary tools. We've provided a list of URLs at the end of the chapter where you can download each tool, as well as get more information.

SECURITY SCANNERS

Scanners are the most commonly encountered kind of security tool. If you've played around with a security tool, chances are it is one of these. Despite their name, these tools frequently do a lot more than just "scan" or look for vulnerabilities. System administrators use them to identify hosts on the network (authorized and unauthorized), to test machines for vulnerabilities to apply patches, and to inspect incoming and outgoing traffic for signs of network intrusion. Hackers use them to scan the network for available services, probe vulnerable machines, and sniff passwords, cookies, and confidential data. Whether intended for a system administrator or test engineer (like SAINT or EtherPeek) or a malicious hacker (like Ettercap), these tools do pretty much the same thing.

Developers and testers can use these tools to find known vulnerabilities in their host system or application platform and to learn about the protocols their application uses. But do not expect that a big investment in a security scanner can fix your application security problems; these scanners are intended for system administrators and end users, not developers. Because they mainly find known problems, for the most part, they might not be helpful in finding new bugs in your product.

Comprehensive Scanning Tools

These tools scan for known vulnerabilities and are developed with the system administrator in mind. Typically, you give these tools an IP address or range of addresses, and they scan each port number successively looking for ones that are open. They also look for version numbers and other specific responses that indicate the software on the remote machine is vulnerable. Alternatively, some perform a more exhaustive examination—checking file contents, permissions, application versions, etc.—but are required to be installed on the local machine.

The most ancient and famous of these tools is *SATAN*, the Security Administrator Tool for Analyzing Networks. Originally, SATAN scanned for 10 vulnerabilities in Unix systems that were popular among hackers at the time it was written. Most of these were misconfigurations in Network File System (NFS) or FTP that allowed an unauthenticated remote user to gain access to the filesystem of a machine. Medium and large sized Unix installations such as universities would often leave these services open to anonymous access rather than deal with the headache of configuring and maintaining a directory service like NIS. SATAN also scanned for versions of the Unix sendmail software that were vulnerable to buffer overruns. The results of these scans would be reported in a Web page produced by the tool. The system administrator would then be expected to fix the vulnerabilities identified by making the appropriate changes to configuration files, upgrading the sendmail version, etc. The hacker who used SATAN would use the results to carry out his exploitation of the system. The modern Unix security scanner *Nessus* works on the same principle as SATAN. However, it provides two features that make it more versatile and maintainable in the modern security environment. First, it implements individual scans in the form of *plug-ins*, with one or more plug-ins for each known vulnerability. As of the time of this writing (February 2005), we know of 2,035 Nessus plug-ins covering 1,504 unique vulnerabilities (source: *www.nessus.org*). Second, it includes NASL2, the Nessus Attack Scripting Language. This C-like language makes it easy to script new scans and vulnerability signatures to create Nessus plug-ins. Some reports of new vulnerabilities include an NASL2 script that can be used to scan for the vulnerability.

Nessus can operate in two modes, a *banner check* mode, in which only signatures are used to determine whether a host application is vulnerable, and *destructive* mode, in which an exploit itself is used to test for vulnerability. The advantage to banner check mode is that it does not compromise a system to determine vulnerability. However, banner checks are subject to both false positives (system is determined to be exploitable when it is not) and false negatives (system is determined to be not exploitable when it is). In fact, some applications and network appliances even manipulate banners to disguise potentially vulnerable services. Destructive mode ensures an accurate test result, but with the side effect of accomplishing whatever malevolent actions are contained in the exploit. Figure 3.1 shows the Nessus Scanner configuration.

Nessus is used primarily with Unix-based systems. For Windows, Microsoft provides its own free comprehensive security scanner, the Baseline Security Analyzer. Microsoft is working to incorporate security scans for all of its major applications into this single tool. Because it is supplied by the software vendor, the tool

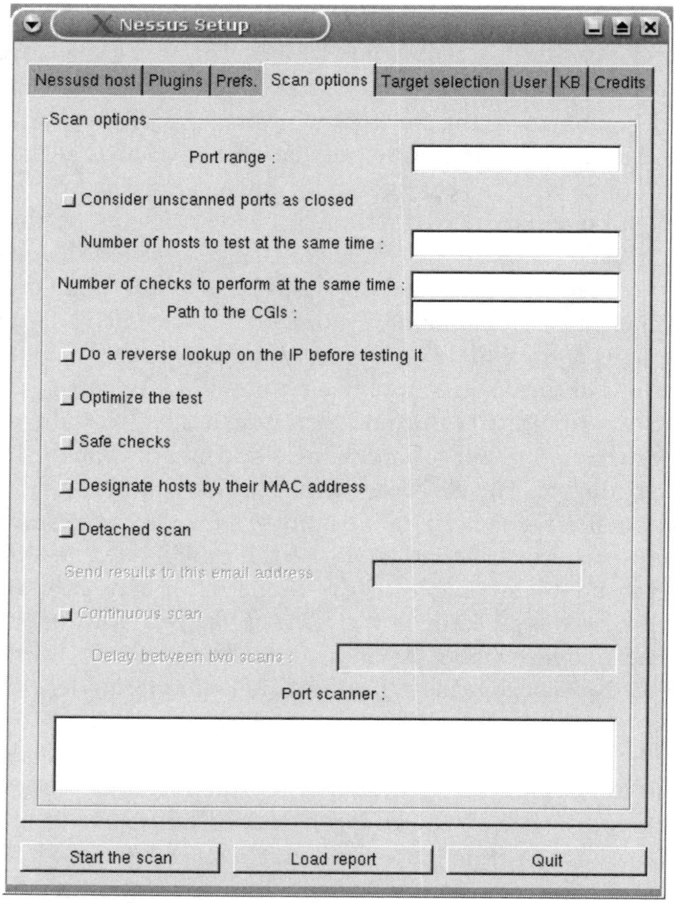

FIGURE 3.1 Nessus Scanner configuration.

has some advantages and disadvantages. It is actively maintained and updated by Microsoft and is integrated with Windows Update, which means it is automatically updated when new vulnerabilities are identified. However, it can find only vulnerabilities for which Microsoft has written a plug-in. This means that, for the most part, Baseline Security Analyzer can't find a vulnerability until after a patch has been released for that vulnerability. Additionally, Microsoft usually has a narrower view of security vulnerabilities than the rest of the software community; so not all published vulnerabilities are identified by Baseline Security Analyzer. Figure 3.2 shows Microsoft's Baseline Security Analyzer.

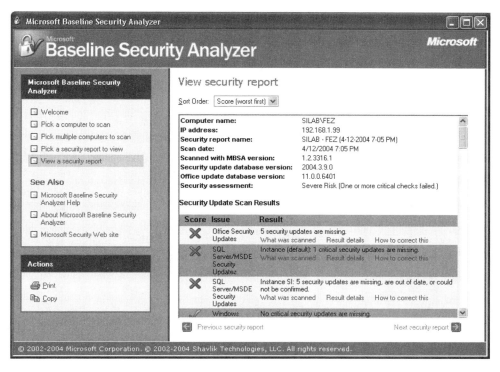

FIGURE 3.2 Microsoft Baseline Security Analyzer.

Nmap and Network Scanners

Network scanners examine a network to identify the hosts and applications available on that network. Rather than just looking for vulnerabilities, these tools help the user gain a better understanding of the whole network: its topology, what hosts make up the network, and what operating systems and applications are running on the network. They essentially perform two functions.

Port scanning identifies all the network applications running on a particular host by attempting to connect to the port number associated with that application. For example, a system that allows an arbitrary connection to TCP port 80 is likely running Web server software. Further inspection of the data returned by the connection can be used to determine what version of the application is running; for example, the response:

```
220 hq.se.fit.edu Microsoft ESMTP MAIL Service, Version:
5.0.2195.6713 ready at Sat, 28 Feb 2004 14:17:20 -0500
```

to a connection on port 25 reveals that the Microsoft ESMTP Service (used by Exchange) is running on this machine. From this one connection, we have not only found out that a mail server is running on the scanned machine, but also that it is likely a Windows server running Exchange.

OS fingerprinting aggregates the data from connections to multiple ports in an attempt to determine the hardware platform and operating system version that correspond to a machine. The previous scan of port 25 revealed that Microsoft Exchange was running; examination of additional ports would reveal that our system is in fact a Windows 2000 Server running IIS and SQL Server as well as Exchange. OS fingerprinting is also possible because, even when banners are spoofed, operating systems respond differently to subtle probing and to packets containing intentional errors. Together with determination of open services, it is usually possible to make a pretty accurate guess of what OS is running on a machine.

The simple network program *ping* uses an ICMP echo request message to determine whether a host is responsive. If the host exists, is reachable, and has a network stack that is responsive to ICMP messages, it replies to an echo request when its IP address is contacted via ping. If it does not respond, the ping program reports a variety of error codes, including "request timed out" and "destination unreachable."

Nmap, an open source tool made available by Fydor of *insecure.org*, is the de facto standard network scanner. Nmap is supported on most platforms and has an enormous number of options related to scanning, OS fingerprinting, IDS evasion, and stealth operation. Nmap is freely available at *www.insecure.org/nmap/nmap_download.html*, where you can select your platform.

Pinging and Port Scanning to Determine Whether a Test Has Crashed the Machine

Network scanners and the ping program can be used to determine whether a host or application service crashed as the result of a test case. If the system stops responding when it ordinarily would (or responds differently), the test case has likely put the system in that non-responsive state. In fact, many commercial security scanners, cracking tools, and protocol testers use this technique to determine whether their exploit or test case was successful. We use both ping and an Nmap SYN scan of the application's port after each test case during protocol testing.

Packet Sniffing and Spoofing

These tools record all traffic associated with a particular host or network. They work because, in an unswitched network, *all* data is transmitted to *all* hosts. Each host must determine which data is intended for it and then pass that data to the correct application. *Sniffers* intercept this data passively by opening the network access layer device (usually the Ethernet card) in *promiscuous mode*, meaning it intercepts all traffic regardless of whether it is destined for the hardware address associated with the device. This allows the sniffer to "see" all of the traffic on the network independent of the operating system's networking stack. The sniffer then reconstructs the data as it would appear to the application, usually with the aid of helper libraries built for that purpose.

Ethereal is the most popular sniffer program with a graphical user interface. It is freely available from *www.ethereal.com/*. Ethereal decodes all of the traffic on the network and presents it in table form, organized by protocol field. According to the Web site, Ethereal contains decoders for 472 protocols including all of the most common ones, meaning it can further decompose application layer data (inside of a TCP stream). This feature is not found in many sniffers. Ethereal also allows filtering of the results based on a rich, object-oriented filter language. For example, the filter expression:

```
ip.dst eq www.mit.edu
```

would display only packets whose IP destination address was *www.mit.edu*.

Tcpdump is a command-line packet sniffer and network analyzer. Tcpdump is the favorite of network programmers, test engineers and serious hackers because it is easily scriptable and can be used remotely and on systems that cannot display a graphical user interface. Many of the tasks that can be performed visually in Ethereal can be performed programmatically through Tcpdump.

Both Ethereal and Tcpdump use the *Libpcap* library for packet capturing and decoding. Libpcap is for Unix but has been ported to Windows (WinPcap) and Java (Jpcap). This library abstracts the task of opening and managing the hardware device and decoding the captured packets. The user can set up a callback function that is called each time a packet is intercepted, and then decode it using the built-in functions. Libpcap is useful for programming custom sniffers in situations where Ethereal or Tcpdump won't do.

Some sniffers can do more than just passively record network traffic. The *Ettercap* sniffer from Alberto Ornaghi and Marco Valleri (*http://ettercap.source forge.net/*) can sniff on switched networks, decode SSL and SSH transaction data, perform man-in-the-middle attacks, and inject data into a stream. These features are essential in doing security testing involving more complicated protocols; however, Ettercap is somewhat more difficult to use than other sniffers and works less reliably. It is better to use it only when you require one of its unique features.

Packet Sniffing to Determine the Results of a Security Test

Network sniffers are extremely useful in functioning as the *oracle* in security test cases. An oracle is a program that determines whether a test passed or failed. The sniffer can be used to create a baseline data set about a protocol; this baseline can be compared against recorded results during testing. If a significant difference is found, the difference can be reported as an anomaly. Sniffers can also make a passive record of the traffic associated with a test session, which can be used in fault isolation and test reproduction. Figure 3.3 shows an Ethereal network analyzer.

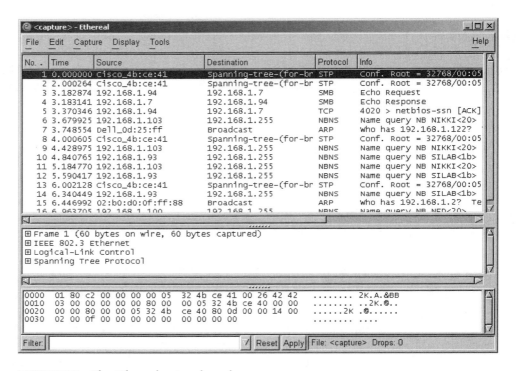

FIGURE 3.3 The Ethereal network analyzer.

HACKING AND CRACKING TOOLS

Unlike network scanners, these tools are designed primarily with hacking in mind. Some of them (packet replay tools and network fuzzing) are useful in finding new vulnerabilities in network-based applications. Web test tools are useful in finding common Web site application vulnerabilities. The primary legal uses of these tools are to educate people on the relative weakness of certain kinds of passwords and to recover your own (or your associate's, if you're the administrator) lost password. These tools help you find novel vulnerabilities in your own application as well as demonstrate how a hacker finds new vulnerabilities.

Password Crackers

Most password-cracking tools rely on brute force dictionary attacks. This kind of attack assumes that the hacker has access to an encrypted or encoded version of the password database and a dictionary of commonly used usernames and passwords. The program systematically encrypts each username and password combination using the same algorithm as the operating system and tests the encrypted result against the password database. When a match is found, the tool has "cracked" one password. This technique is successful because modern computers can try many username and password combinations in a relatively short period of time, and many systems have at least one common username. For example, almost all Windows computers have a user named "Administrator." Passwords that are made up of relatively short dictionary words or personal names are easiest for a password cracker to crack; exceptionally long passwords (13 letters or more) and passwords with non-alphanumeric characters are especially difficult to crack.

John the Ripper is a relatively simple and commonly used password cracker for Unix. John relies on the fact that the Unix `crypt()` function, used to encrypt and decrypt passwords, can be called by any program. The format of a Unix password database entry is:

```
secret:$1$lYyNg9Do$nW7FpiK7o3Bof4.dwJ8dn/:12477:0:99999:7:::
```

John can read the encrypted password string (beginning `$1$` and ending `dn/`) and compare it to the results returned by a call to `crypt()` with a dictionary password. If the strings match, the password is cracked. Because `crypt()` is callable directly, no failed authentication attempt is recorded in the log.

L0phtcrack (now LC4) from @Stake, Inc., performs a similar function on Windows. L0phtcrack is able to deal with the idiosyncrasies of the Windows SAM password database format, which is more complicated than Unix's.

Packet Generation and Replay

In security testing, it is often necessary to write packet data directly onto the wire. It is certainly possible to open up a socket in your favorite programming language to accomplish this, and in many cases, that's the best way. However, lots of tools exist to help with packet generation and replay. Some are able to open more diverse kinds of connections, or connections at a lower level, than a high-level language networking library like BSD Sockets or Winsock. Others are designed to make packet injection easier for people without networking knowledge.

The simplest of these programs, Chris Wysopal's *Netcat*, (*www.atstake.com/research/tools/*) functions similar to the `cat` command in Unix or `type` command in Windows. Netcat reads data from standard in, which can be piped from a file, and writes it out on a network socket. It can be used to send arbitrary data to a TCP or UDP application listening on a particular port. The command:

```
echo -en "GET / HTTP/1.0\r\n\r\n" | nc www.myserver.com 80 > page.html
```

retrieves the home page of *www.myserver.com* and stores it in the file `page.html`. When used in *listen* mode, Netcat can bind to a specific port and output any data written to that port by a remote host. This allows the user to simulate a remote server when testing a client application.

Nemesis, and the library it is based on, *libnet*, enables you to create packets that are not easily created using an operating system's TCP/IP stack. For example, it is not easy to send "spoofed" packets, packets that appear to come from a different IP than the sending machine, using a conventional BSD or Winsock socket. Nemesis and libnet can overcome this; they are able to write to the network access layer device (Ethernet card) in the same way that a sniffer can listen to the device. As a result, they can send fragmented packets, packets with corrupt headers, packets with invalid checksums; in short, anything can make it to the live wire. They are especially useful in testing router and network appliance implementations, which have to deal with a broader set of protocols and tolerances than an end-user system.

Two companion tools to Tcpdump useful in generating and sending packets are *tcpslice* and *tcpreplay*. tcpslice is used to cut, paste, and reassemble packet sequences captured with a sniffer like Ethereal or Tcpdump. It is possible to extract one particular TCP session, or multiple sessions, from out of a sniffer capture file based on port number, source address, etc. tcpreplay allows captured data to be *played back onto the wire*, with or without timing adjustments. The problem with tcpreplay is that it plays the data back without correcting IP checksums, and without correcting timestamps that appear in some application layer protocols, so the remote host you are testing might not respond appropriately to the replayed data.

The teardrop attack, a popular denial-of-service attack used by hackers in 1998, resulted from sending an overlapping set of fragmented IP packets. When the networking stack of Windows and Linux machines tried to reassemble the fragments, a pointer error caused the stack to crash irrecoverably. This packet sequence, which could not be created by a normal `socket()` *call, was trivially easy to generate and test with libnet.*

Network Fuzzing

Fuzzing involves sending random data to a network port or application in an attempt to find buffer overruns and denial-of-service vulnerabilities from incorrect parsing of the data. Fuzzing is useful in binary protocols and any other protocols that lack documentation. It is also useful in forcing exception handling code within an application to be invoked that could not be easily reached otherwise. One way of doing this is to use the `urandom` device in Linux to generate some random data, and Netcat to send it. The command:

```
cat /dev/urandom | nc 192.168.1.1 80
```

accomplishes this nicely.

Spike, written by Dave Aitel of Immunity, Inc., (*www.immunitysec.com/*) is a fuzzer creation kit. It allows you to create a fuzzer customized to the protocol you are studying, handles checksums, and length field markers and is capable of inserting interesting "fuzz strings." These strings include directory traversal attack strings (e.g., ../../../../../../../../), long strings that can be used to find buffer overruns, and names of system files (e.g., /etc/passwd).

Web Site Test Tools

Because they are designed to permit users from anywhere on the Internet, Web-based applications are particularly vulnerable to attack. In addition, the stateless nature of Web pages, heavy use of server- and client-side scripting, and the emergence of application server software have created some security issues unique to the Web application paradigm. Some attacks, such as cross-site scripting, are unique to the Web. Others, like SQL injection or OS command injection, are occasionally found elsewhere, but are common in Web applications. SANS (*www.sans.org*) maintains a "top ten" list of these bugs at their Web site. A similar list for Web applications can be found at OWASP, the Open Web Application Security Project (*www.owasp.org*).

Nikto, an open source Web site scanner, is capable of finding these vulnerabilities. In addition to the "top ten" Web site bugs, Nikto can find version specific vulnerabilities in major application server platforms, unsafe CGI and server-side scripts, and known vulnerabilities in major Web servers. Nikto works by "spidering" a Web site—successively following links from one page to the next, performing its tests on each page.

REVERSE ENGINEERING TOOLS

Security testing requires an understanding of the inner workings of a program. If you're the sole developer of an application, or have ready access to *understandable* source code, these tools won't help very much. If you don't, however, reverse engineering tools can help you learn more about the application you're testing.

As we mentioned in Chapter 2, some legal issues surround reverse engineering. You should be conscious of these if you decide to employ one of the tools in this section.

Source and Binary Scanners

Many security vulnerabilities are the result of programming errors that are easily detectable by inspecting the source code. Programmers might use unsafe versions of library functions, such as strcpy or strcat, and security objects like encryption keys might be freed in memory without overwriting their contents. Source scanners attempt to identify these potential vulnerabilities by searching a source file for the signature of the programming error.

RATS, the rough auditing tool for security, is a source scanner for C, C++, Perl, PHP, and Python. RATS contains signatures of 484 vulnerable functions and other programming errors. However, the algorithm RATS uses to scan code is not very sophisticated, and it cannot find certain categories of vulnerabilities like signed/unsigned bugs.

WHY NOT JUST SCAN EVERYTHING?

Source scanners and binary scanners are part of a family of tools called *static analyzers*. Static analyzers attempt to make assertions about a program without observing the program in a running state. As a result, they are good at identifying *potential* vulnerabilities, but not in determining whether those vulnerabilities actually exist in the running program.

Consider the following piece of C code:

```
char* newStr(char* oldStr)
{   char* tmp;
    tmp = (char*) malloc(strlen(oldStr) + 1);
    if(!tmp) return NULL;
    strcpy(tmp, oldStr);
    return tmp;
}
```

This function creates a new dynamic character array and copies the string oldStr into it. A single line source scanner would report the strcpy as unsafe, even though tmp is allocated with the same number of bytes as oldStr.

> This results in a very large number of false positives. In programs with more than 10,000 lines of code, it is not generally possible to test every potential vulnerability reported by a source scanner. This problem is compounded in the case of binary scanners, as it is even more difficult for a binary scanner to "look back" to previous lines to sort out buffer sizes and data sources.

BugScan, developed by HBGary (*www.hbgary.com*) but recently acquired by LogicLibrary (*www.bugscan.net/*) is a commercial tool that scans binaries (executable programs and DLLs) to find vulnerabilities. While we have not used BugScan, the tool looks promising. Binary scanning has an advantage over source scanning in situations where you do not have access to the source code, or it is written in a language that is not supported by available source scanners. However, binary scanning programs have even less information to work with than source scanners.

Specialty Editors

A variety of tools exist to help in decoding proprietary file formats, including the executable file formats of programs. Simplest among these are hex editors like *WinHex*. WinHex allows you to edit files in ASCII, Unicode, hexadecimal, and binary format, as well as directly edit disk sectors and memory pages. It also accommodates searches in multiple data formats, which is useful in backtracking from debugger to the data's source within a file.

PE file editors allow you to directly edit files of the Windows EXE (portable executable) file format. Many of these editors also allow you to edit and manage exports (functions exported for use by other programs), imports from DLLs, and Windows resources embedded within these files. Heaventools' *PE Explorer* is a PE file editor that supports all of these features.

The freeware tool *Resource Hacker* by Angus Johnson (*www.users.on.net/johnson/resourcehacker/*) can inspect, modify, and delete Windows resources within an EXE or DLL file. Many applications store their strings, dialog, and Web page information in resources; inspecting and modifying these aids in reverse engineering.

API and System Monitors

API and system monitors are used to inspect the "boundary" between an application and the operating system that surrounds it. The information that is passed on this boundary is very useful; with few exceptions an application cannot open a file, connect to the network, or create a child process without making an operating system call. In fact, most any data that is passed into or out of an application involves a system call of some kind. Many system calls, when called incorrectly, are as unsafe as strcpy or sprintf. API monitoring tools allow you to watch an application making those calls and inspect the data it passes.

strace for Unix prints out the names and parameters of all system calls made by an application. These include file, network, and process calls. strace can do this because most system calls are made by calling a kernel function with the system call number; a single dispatcher is used to intercept and handle system call events. strace intercepts all calls within the kernel at this location. strace does not intercept library calls, so common functions like printf, which is part of libc, are seen by strace as a write() call to the console device.

Sysinternals.com publishes a number of freeware utilities for API monitoring in Windows. These tools have self-explanatory names: *Diskmon*, *Regmon*, *Filemon*, *Pmon*, and *Portmon*. The Sysinternals tools have a graphical interface and are more concrete than strace, because many Windows programs call the Win32 API functions directly. They are invaluable in finding a number of vulnerabilities involving temporary files, data stored in the registry, and undocumented network protocols. Figure 3.4 shows Regmon.

Disassemblers

Disassemblers are programs that translate binary executable files into their assembly source code. While they do not recover the original code as it was written, they make it easier to view some of the structure and algorithms that make up the program. *IDA*, the interactive disassembler from DataRescue, Inc., is a very powerful disassembler that supports multiple architectures and file formats. IDA is able to detect the source language and compiler using code profiling, does code flow analysis to determine relationships among blocks, and is capable of producing call flow graphs of the program. It also supports custom-written plug-ins—many folks in the reverse engineering community rely on IDA to do the upfront work of disassembly and code flow analysis and implement their proprietary tools as IDA plug-ins.

Using Debuggers for Security Testing

Conventional software debuggers have a number of unconventional uses in security testing and can help quite a bit with program understanding. Because they interact with the program while it's running, you can follow data as it is passed from one function to another, search the memory space for values such as passwords and encryption keys, and observe the effect of fuzzing data on a program. You can even

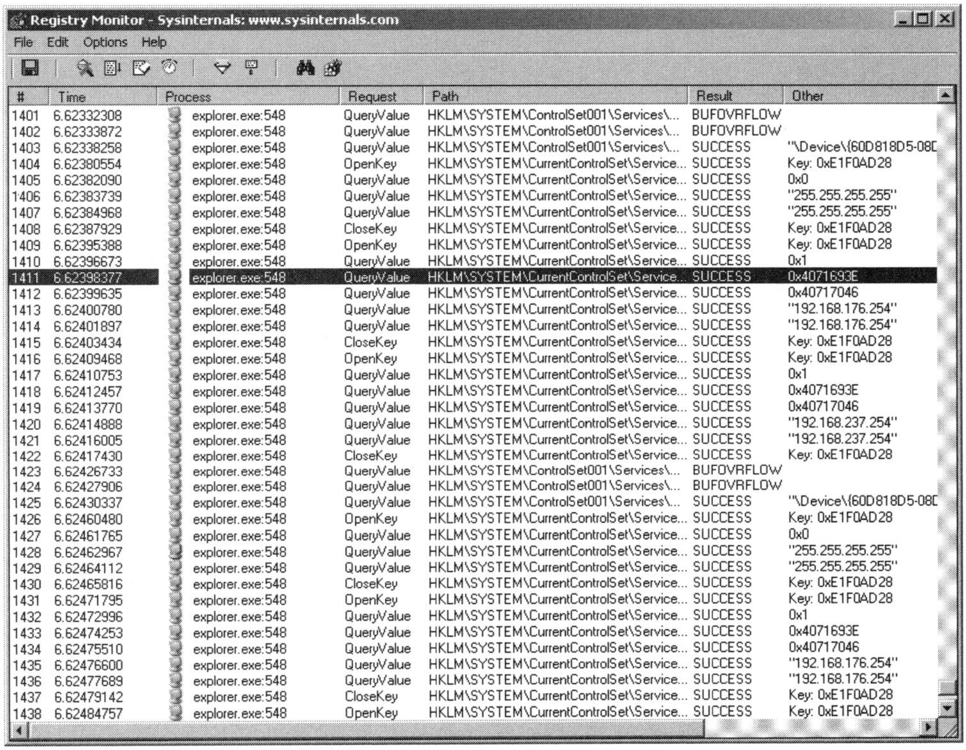

FIGURE 3.4 Regmon.

use a debugger as an "oracle" for file and network corruption testing. We take advantage of the fact that a program running in a debugger traps into the debugger when an exception is raised by using this fact to know when our corrupted data has "crashed" the program we're testing. By looking at the values of registers and the stack, we can determine whether this crash is exploitable.

Our favorite debugger for Windows is the shareware *OllyDbg*. OllyDbg is an assembly level debugger, so you won't be able to single-step source code if you have it. However, its exception management, use of the processor debug registers, data breakpoints, and plug-ins mean it can work around standard anti-debugging tricks used by some programs. OllyDbg is shown in Figure 3.5.

The NT Symbolic Debugger (NTSD) is bundled with Windows NT, 2000, and Windows XP. While NTSD does not have the pretty interface of OllyDbg, it is the debugger Microsoft uses internally on most application projects and as a result is

52   The Software Vulnerability Guide

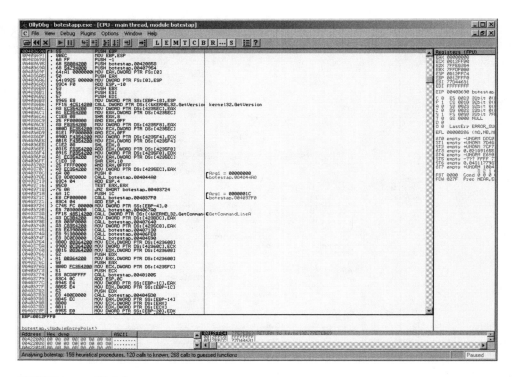

FIGURE 3.5   OllyDdg.

able to debug Windows applications with less difficulty than other debuggers. We especially like the command-line interface and advanced features, including handling of crash dumps and Dr. Watson info, as well as its capability to debug services.

Another favorite of the reverse engineering community is Compuware's *SoftICE*. SoftICE is a system level debugger originally written for use in device driver testing. It is useful in debugging kernel functions, drivers, and other components that are not ordinarily accessible in "user" space. However, SoftICE's system level perspective means you're debugging the *whole computer* at once, which can be cumbersome when you're only concerned with one application.

For Linux, there really is no better debugger than GDB. GDB can work as both a source and system level debugger, and is highly scriptable. Its only drawback is its awkward, command-line interface, necessitated by the number of platforms it supports.

COMMERCIAL TOOLS

Most of the tools we have presented in this chapter are free for the taking; the tools in this section are not. They do, however, represent reasonable attempts by their respective vendors to make high-quality security testing tools for developers. Part of the appeal of these tools is support; the vendors are experts in the use of them, and security in general. Most offer training and maintenance contracts that ensure the tool is properly integrated into your development cycle. They are well worth the price if your company is making an investment in security testing.

Retina

Retina is a comprehensive security scanner like Nessus. Its main advantages are that it is developed and supported by the eEye Digital Security Team, who keep the tool very up to date with the latest vulnerabilities, and its speed compared to other scanners. It is also relatively inexpensive and easy to use if you have only a small network to scan. Licensing is on a per-IP address basis.

AppScan

AppScan is a Web security scanner based on the AppShield technology from Sanctum, Inc. AppScan spiders a site and finds OWASP top ten vulnerabilities. It has a very well-designed user interface based on a custom Web browser. It injects faults into the underlying HTTP traffic destined for the Web server and then compares the results page to a page with no injected vulnerability as well as known error pages.

WebProxy

WebProxy from @stake, Inc., is a different kind of Web testing tool that can find OWASP vulnerabilities as well as common Web and application server bugs. Rather than working as a browser/spider, it actually acts as a proxy server between the client and the Web server. This means hidden fields, cookies, authentication tokens, and other Web page elements can be modified in the middle of a transaction. Because it includes a "fuzzing" component, it is also capable of finding buffer overflows.

Holodeck

Holodeck is an API level fault injection and analysis tool offered by Security Innovation®. It goes beyond traditional API monitoring tools in its capability to *inject faults* into the application it is testing. Holodeck can simulate latent error conditions that occur in software, such as low memory conditions, disk failures, and loss of network connectivity, and measure the application's response to those failures. It can also inject user-created faults into the application, both through the user interface and a COM API. The Holodeck main window is illustrated in Figure 3.6.

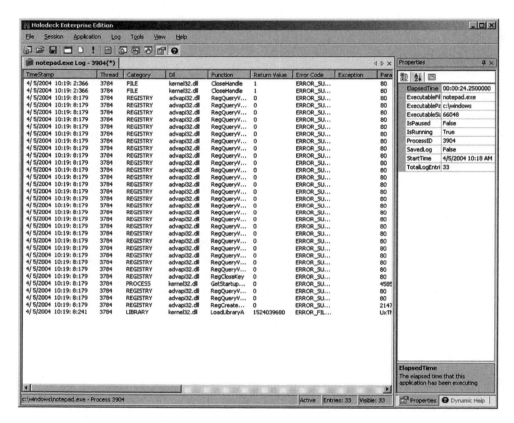

FIGURE 3.6 Holodeck main window.

Don't be discouraged if you don't understand everything about each of these tools from what is written in this chapter. Tools are only part of what you need to find security vulnerabilities. The remainder of this book describes techniques for finding specific kinds of vulnerabilities using these tools. The "For More Information" section has links to the documentation and downloads for each of these tools.

FOR MORE INFORMATION

- SATAN: *www.fish.com/satan*
- Nessus: *www.nessus.org*
- Microsoft Baseline Security Analyzer: *www.microsoft.com/technet/treeview/default.asp?url=/technet/security/tools/mbsahome.asp*
- Nmap: *www.insecure.org/nmap/index.html*
- Ethereal: *www.ethereal.com*
- Tcpdump and Libpcap: *www.tcpdump.org*
- Ettercap: *ettercap.sourceforge.net/*
- John the Ripper: *www.openwall.com/john/*
- L0phtcrack, Netcat, and WebProxy: *www.atstake.com*
- Spike: *www.immunitysec.com/resources-freesoftware.shtml*
- Nikto: *www.cirt.net/code/nikto.shtml*
- RATS: *www.securesoftware.com/resources/tools.html*
- BugScan: *www.bugscan.net/*
- WinHex: *www.x-ways.net/winhex/index-m.html*
- PE Explorer: *www.heaventools.com*
- Resource Hacker: *www.users.on.net/johnson/resourcehacker/*
- Filemon, Regmon, etc.: *www.sysinternals.com*
- IDA: *www.datarescue.com*
- OllyDbg: *home.t-online.de/home/Ollydbg/*
- SoftICE: *www.compuware.com*
- Retina: *www.eeye.com*
- AppScan: *www.sanctuminc.com*
- Holodeck: *www.securityinnovation.com/holodeck/*
- Open Web Application Security Project: *www.owasp.org*

Part II

System-Level Attacks

4 Problems with Permissions

In This Chapter

- The Bell-Lapadula Model
- Description
- Fixing This Vulnerability
- References

Most traditional writing about computer security, especially works prior to the popularization of the Internet in the late 1990s, was concerned primarily with access control, and for a good reason. When computers first came about, we had no need for security in a software sense; the first major computers were large enough to require dedicated rooms, and access was secured by some physical means—a door lock, a premise alarm, etc. The advent of multi-user systems led to a need to protect users from each other, and protect the system so that it could remain a shared resource. Originally we had no concept of protection from remote attack, because typically no remote users existed. (Occasionally a modem left attached by a system administrator facilitated a remote attack.) As a result, security was primarily about limiting the access of a user to only those resources appropriate to him. Numerous models have been developed based on geometric concepts: the ring model, the hierarchical model, the superuser model, etc. They all follow the same basic principles, though. These principles are as follows:

- A system is made up of a number of *objects* or resources, and a number of *subjects* or users. Most typically these objects are files and directories, though other objects, such as pages of memory, devices, pipes, and semaphores might also be subject to access control.
- Each of these objects has a number of access modes, typically, *read, write*, and *execute*. Some systems include other access modes such as *create* and *delete*, while others treat these as the same as a write.
- Users have permission to access an object in one or more of these modes through a permission scheme.

In addition, most permission schemes are hierarchical, meaning some users have more access rights than others, and these users can typically access more resources than the others. Some schemes have a concept of *ownership*, in which the primary rights of access rest with one particular user, who can grant or revoke these rights from others.

THE BELL-LAPADULA MODEL

The Bell-Lapadula model [Bell73] is one of the older and more recognized models for implementing access control. It is based on the hierarchy used to protect classified information within the U.S. government and is organized as follows:

- It has multiple, ordered security levels: top secret, secret, confidential, and unclassified.
- Each object is assigned a security level, as is each subject.
- A subject can read or write objects at his level. For example, a subject with a top secret level can create, read, or write files with a level of top secret.
- A subject can "read down." This means a subject can read, but not write, files at a level below his. For example, a top secret subject can read any secret or unclassified objects within a system.
- A subject can "write up," meaning create and write to objects at a level higher than his. This is because anyone with a higher level has permission to read it anyway.

Though it is the model most frequently taught regarding access control, in practice, the Bell-Lapadula model poses some significant problems. First, you have the issue of *privacy*. Any user with top secret credentials can read any file at the top secret level; you have no means of segregating files among users. Second, no well-defined mechanism exists for changing the level associated with objects. Because no user can "write down," you cannot downgrade the level of a file. Figure 4.1 illustrates the Bell-Lapadula model.

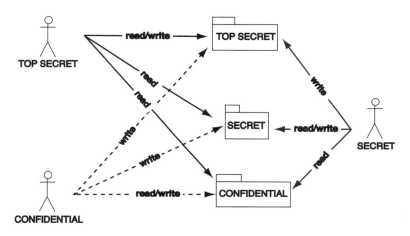

FIGURE 4.1 The Bell-Lapadula model.

As a result of these limitations, operating systems cannot simply implement the Bell-Lapadula model. In practice, their schemes can be relatively simple (systems that permit only one "superuser" such as the "root" user in classic Unix implementations have only two kinds of privilege, ordinary user and superuser, to worry about) or quite elaborate (systems that support access control lists, or ACLs, such as Windows NT, allow many functions within the operating system to be assigned multiple levels of privilege based on defined roles). The ACL approach is strongly preferred by security professionals, because it allows the administrator to have more fine-grained control over individual privileges than the superuser model, but both approaches have drawbacks. In the superuser model, because the privilege of writing to arbitrary files or adding and removing users is reserved for the superuser, applications that need to make use of those functions must be run as superuser. As a result, that application has all the *other* privileges that a superuser has, in addition to those it really needs. Misconfiguration of the application might result in a user being able to access some of those additional privileges. On the other hand, the ACL approach requires the application developer or system administrator to be familiar with, and correctly define, the appropriate roles for each action an application wants to perform. The more "fine-grained" the access controls can be, the more likely it is that a developer or end user can make a mistake that exposes a vulnerability. Complexity sometimes is the enemy of security.

This chapter deals with several issues surrounding multiprivilege systems:

- *Permission misconfigurations* occur when an application ships with the default permissions set incorrectly, allowing an attacker to use the application to circumvent security.
- *Escalation of privilege bugs,* sometimes called *suid bugs,* in which an application running as a privileged user can be exploited to escalate privilege, is found primarily in Unix, but can also occur in other operating systems.
- *De-escalation bugs* take advantage of shared special purpose accounts designed to avoid the first two categories of bugs. For example, a user who "de-escalates" from himself to "nobody" might be able to kill processes running as nobody, a problem in shared environments.

DESCRIPTION

Support for multiple users was one of the "killer apps" that made Unix a mainstream operating system during the 1970s and 1980s. While other systems could be used by only one user at a time or required a dedicated CPU for each simultaneous user, Unix's timesharing meant that CPU resources could be alternated between multiple users, giving the appearance that it could perform multiple simultaneous operations. Unix borrowed its user concept from many other systems of the day including MULTICS, and the same basic concept underlies modern Unix variants such as Solaris or Linux. Each user in Unix has an account that is accessed by means of a username and password. All resources within the system have a corresponding *ownership*; that is, they belong to one and only one user, who can assign permissions to others to use it. The Unix permission scheme is implemented by means of a permissions mode number, with each bit in the mode corresponding to a permission assigned or denied to a particular category of user. The exact permissions vary slightly from one Unix implementation to another, but the basic idea is as follows:

- You have three categories of users: the owner, the group owner associated with the resource, and "everyone." Permission numbers are made up of three digits, one each for owner, group, and everyone. In the permission number of 654,

6 corresponds to the owner, 5 to the group, and 4 to everyone. Historically, group permissions were not used extensively in Unix. Some modern applications, such as MySQL, rely on them extensively, though.
- You have three kinds of permission: read, write, and execute. These values form a 3-bit mode as shown in Table 4.1.

TABLE 4.1 Permission Mode Bits in Unix

Bit	Meaning
100	User has read permission to the resource.
010	User has write permission to the resource.
001	User has execute permission to the resource.

For example, a resource with a permission mode of 755 would mean the following:

- Owner has read, write, and execute permission (all three bits are 1).
- Group has read and execute permission, but not write permission (first and third bits are 1, but second bit is 0).
- Everyone has read and execute permission (same as previous).

In addition to these ordinary permissions, a special flag called the *supervisor bit* can be set on a file. This bit tells the operating system to execute the program as though it were being executed by the owner, with all the permissions associated with that user. The operating system uses the setuid() family of functions to do this. Many operating system commands and services have the supervisor bit set so that they can be run as an ordinary user but perform a function that requires "root" access to accomplish a task. For example, the passwd command needs to update the password file, owned by "root," in order to successfully change a user's password. Likewise, several applications within X Windows deal with logging in and out as different users; the X server and some of these applications need to have the supervisor bit set.

Remember, because the user ID of the process is effectively changed to that of the owner when the program is executed, whatever actions can be performed by the program can be performed as the owner. So a program owned by root with the supervisor bit set could conceivably do anything that root could do. This is because, as we explained previously, the access control model of Unix is too simple to differentiate between a superuser who needs to perform a specific task and a general purpose administrator. For example, it is not, generally speaking, possible to give a user only rights to add and delete users or change passwords without giving him all other rights to the system.

An attacker can exploit this model a number of ways. One way would be to find a piece of side-effect functionality within the program that performs the task he wants. For example, an attacker who knew that the program deleted a file might be able to use that functionality to delete an arbitrary file. Alternately, the program might fail to process commands or input correctly, meaning that the attacker can pass arbitrary commands to the program. Finally, an attacker might be able to find a buffer overrun in the program, allowing the arbitrary code to be executed as the superuser.

It should be noted that while the preceding description is generally true of Unix systems, a considerable amount of variation exists from system to system. For example, Linux is presently implementing Posix capabilities as a mechanism for providing more fine-grained access control.

Finding Programs with the Supervisor Bit Set

Finding out whether a program has the supervisor bit set is relatively easy. The -l option to ls lists additional details about a file, including its ownership, permissions, and the date it was created. The first 10 columns of each line is the permission mode; an s in the fourth column means the program has the supervisor bit set. Figure 4.2 shows the programs in the /usr/bin directory of Linux that have the supervisor bit set.

Searching for supervisor-enabled programs on the system is just as easy. The regular expression ^...s matches all lines that have an s in the fourth column, so we could search the whole file system for supervisor-enabled files with the command:

```
find –perm +4000 -print
```

Attacking Supervisor Mode Programs by Finding Side-Effect Functionality

Many supervisor-enabled programs exist to bypass conventional operating system security measures. When implemented correctly, they perform only the function specified and guard against potential exploitation by carefully processing all input and anticipating potential misuse. The side effects of specific functions within the program might leave the software open for vulnerability.

Consider a vulnerability that existed in many FTP clients, including the default clients packaged in Linux and Solaris. These clients had an undocumented feature

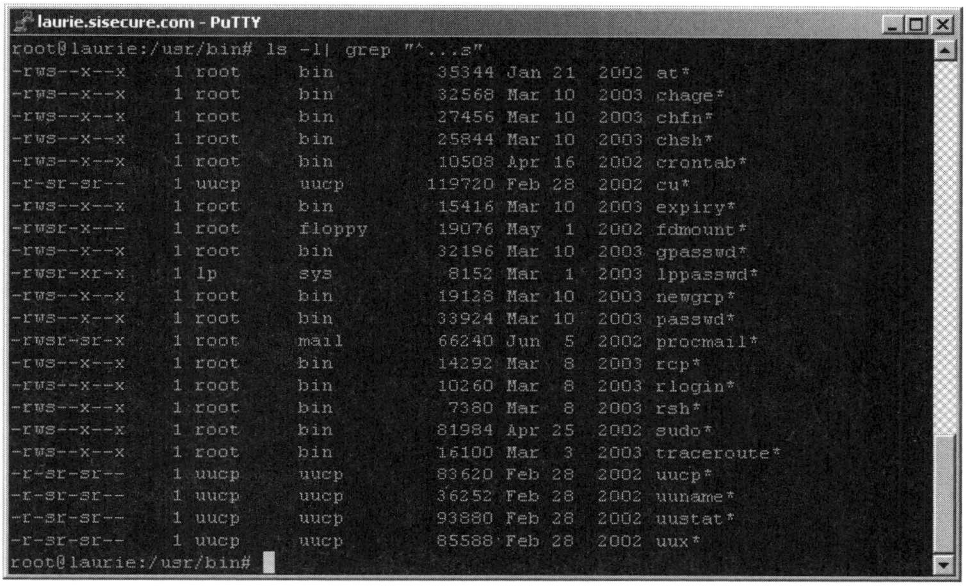

FIGURE 4.2 Programs with supervisor bit set in /usr/bin.

that would allow the contents of a downloaded file to be filtered through a local program. The command:

    get foo |more

would retrieve the contents of the file foo from the remote server and display it using the more (pager) program. The programmers of these FTP clients failed to anticipate the side effect of this feature when coupled with another feature: default local filenames. This feature allows a user to omit the name by which the retrieve file is saved; the client defaults to saving the file with the same name as the remote file. The result is that when the command:

    get |sh

is executed in the client, the contents of a remote file named |sh are retrieved and then passed to the sh program as input. As a result, a malicious FTP server could use this technique to execute arbitrary commands on a client.

This vulnerability existed because the programmers failed to anticipate that files with a pipe (|) as the first character in the name could be placed on the server, and what the side effect of such a filename would be. Pipes and other forms of redirection are a common cause of escalation of privilege vulnerabilities. A very similar vulnerability existed in the Apache Web server running under Windows. When a normal CGI program is executed in Apache, the executable is invoked directly by the Web server with no intermediary command interpreter. However, to support CGIs that were written in DOS batch file language, special functionality was added to invoke cmd.exe (the command interpreter) when a CGI with a .BAT or .CMD extension was encountered. CMD was passed the batch file as an argument; subsequent arguments processed off the request URL were also passed to CMD so that they could be interpreted by the batch file.

When a request URL contained a pipe as the first character in the first parameter, interesting side-effect functionality in cmd.exe was invoked. cmd interpreted the pipe as a redirection command, just as it would if it were typed on the command line. Consider the effect of this command, supplied as an example by Ory Segal in his description of the bug:

```
http://TARGET/cgi-bin/test-cgi.bat?|echo+Foobar+>>
+..\htdocs\index.html
```

When cmd is passed the string |echo as its first argument, rather than passing it along to the test-cgi.bat program, it invokes test-cgi.bat with no arguments and pipes the result to the echo command. The remainder of the arguments define the behavior of echo; in this case, it appends the string Foobar to the root index.html document. Such a technique could easily be used to overwrite the .htaccess and .htpasswd files or execute any other command on the server.

Environment variables are another source of privilege escalation vulnerabilities. Consider a program that relies on a script to do setup or configuration for it, as is the case with many X Windows programs. The script might be read-only, meaning the attacker cannot modify it. However, if the script processes environment variables in an unsafe manner, he might still be able to execute arbitrary commands. Say, for example, an application uses a shell script to delete a temporary directory within a user's home directory. The user's home directory can be safely obtained by searching /etc/passwd for the line corresponding to the username. However, the HOME environment variable is set to the directory name and is more convenient to use within a shell script. The shell script line to delete the temporary directory might look like this:

```
rm -rf $HOME/.temporary_directory
```

While this might look innocuous, it is actually a very dangerous way of employing a temporary directory. The reason is that the user can control the value of the HOME environment variable. Because of this, an attacker can specify another user's home directory as the location of the temporary directory. However, a more serious side effect of this is that rm can remove more than one file with a single command. Files are separated with a space. If the attacker sets his HOME environment variable to be /, the script deletes the entire contents of the hard drive.

> **THE dtappgather VULNERABILITY**
>
> This problem arose from a poor implementation choice in creating a temporary directory associated with the generic display device in Solaris' Common Desktop Environment 1.0.2, found in Solaris 2.5. CDE needed to change the permissions on this directory so that its ownership corresponded to the user that was presently logged in on the X console. To do this, it used the Unix chown() function, which changes file ownership from one user to another. Because chown() can be called only by root, the application that performed this function, dtappgather, had the suid bit set.
>
> The permissions on the directory /var/dt/appconfig/appmanager, where the file was located, were 777, the mode for universal access. The result was that a local user could create a symbolic link from generic-display-0 to a file of his choice; for example, /etc/passwd. When dtappgather was invoked (it was not necessary to log in on the console to do this), the file symbolically linked to generic-display-0 had its ownership changed.

Attacking Supervisor Mode Programs by Exploiting a Buffer Overrun

Implicit side-effect behavior is not the only way to exploit a supervisor mode program. Certain kinds of bugs that result from improper processing of input, including buffer overflows, can also lead to privilege escalation in supervisor mode programs. In this case, an attacker could take advantage of a buffer overflow to execute arbitrary code as the privileged user. The traditional use of these vulnerabilities is to give the user a "root shell," the equivalent of the shell prompt a user would get when logging in as root. This is accomplished by forcing the app to execute code that invokes the shell interpreter, /bin/sh. Because the attacker is already logged into the system as an ordinary user, he can type commands in the shell as though he were the root user.

For a while, both Linux and Solaris were full of vulnerabilities of this kind. They were especially popular in university environments, where shell accounts on Unix machines were relatively easy to come by. Some high-profile ones include:

- `fingerd`
- `dip`
- `mount`
- `uucp`
- `admintool`
- `lpr`
- `lpstat`
- `ld.so`
- `fontfile`

Chapter 8 deals with finding these vulnerabilities in more detail.

Windows: Not Immune From, but Less Prone to, Escalation of Privilege

The bulk of this chapter has concentrated on escalation of privilege in Unix. While privilege vulnerabilities do exist under Windows, they occur less frequently. Several reasons for this exist. First, Windows uses role-based access control, rather than a supervisor model. This means that a user or application can be assigned elevated privileges to access several resources without giving arbitrary access to the system. The operating system itself makes extensive use of ACLs, even for non-file resources such as registry keys, pipes, and interprocess communication channels. It also provides a rich set of roles (called Security IDs or SIDs) ranging from "Administrators" to "Everyone," including special roles for guests, built-in users, power users, services, and anonymous users. The result is that programs do not have to escalate all the way to "Administrator" or "System" to accomplish a task.

How much difference the role-based access controls make in Windows is not really clear, though. First, Windows provides backward compatibility, both in code and philosophy, to the Windows 95 family of operating systems. These systems had no multi-user concept, and programs written for them could create files in any directory, add and delete registry keys, and perform a variety of "unsafe" functions. To preserve compatibility with these programs, the ordinary user must be logged in as "Administrator" even when he is not performing administrative tasks. In fact, on many versions Windows automatically creates an administrator-equivalent account in the primary user's name; this user is automatically logged in when the system boots. So, in practice, most users are *already* "administrator" when they sit down at the machine. For this reason, attackers would not benefit by focusing on finding vulnerabilities that escalate local privilege in Windows. Second, Windows is primarily deployed in two configurations: single-user desktop and standalone server. This "client-server" model negates the need for local privilege escalation. On his own machine, the attacker already has full access. On the central server, where information is stored and shared with other users, he is not "local"; a local privilege exploit would do no good.

We are not suggesting that one system (Unix or Windows) should be selected over the other solely on the basis of access controls. Part of the mess we're in today in security is the failure of security architects to see beyond access controls when these products were created in the 1980s and early 1990s. However, support for ACLs is one of the bright spots of Windows security.

FIXING THIS VULNERABILITY

The best solution to this problem is to avoid the supervisor bit completely. Many services that previously ran as "root" now run as "nobody"; this account potentially provides a higher level of security because the system can restrict some actions from being performed using the "nobody" account, including changing the account's password, logging in for an interactive session, and spawning a shell. If "root" access is required, follow these guidelines:

- Remove the supervisor bit from programs that don't need it. This can be done by chmod a-s <program-name>.
- Limit the functionality of the supervisor-set program to the function it is intended to perform. If a program needs to be suid solely to copy a file or change permissions, consider isolating that functionality into a separate executable and setting the suid bit only on that program.
- Avoid writing suid programs that can write to arbitrary files or spawn shell programs.
- Don't trust any data, especially from the non-suid portions of your own program. These can be manipulated by the attacker. Avoid arbitrary length strings, complicated data structures, and environment variables.
- Avoid assigning the suid bit for convenience or for superfluous reasons—e.g., to directly manipulate the video pages or sound device. Controlled APIs such as DirectX and OpenGL exist to do this.

The setuid() and seteuid() System Calls

The setuid() function allows a suid process to change the user ID associated with that process, effectively dropping privilege. This technique is used by programs like getty that allow a user to log in and launch a shell. Once setuid() is called, the program cannot "go back" to having "root" privileges. This side effect is convenient for services that might need to be started as "root" at boot time, but for which the full privilege set is undesirable. The setuid() function allows such a program to "drop down" to "nobody," where escalation of privilege is not as serious a concern.

The `seteuid()` function (Linux only) allows a program to set its effective user ID. Only a program with "root" privileges can call this function; doing so temporarily switches the user ID of the program from the privileged user to an unprivileged user. Thus, a program that needs suid permissions for a small piece of functionality can "switch" right away back to the user ID of the user who invoked it. Because the program controls at what points it has effective root permissions, a buffer overflow or similar attack would have to occur in the critical (`euid=root`) portion of the program in order to escalate privilege.

DE-ESCALATION: KILLING THE WEB SERVER

An odd side effect of the de-escalation technique is that child processes running as "nobody" can be killed by other processes running as "nobody." Server-side includes in the Apache Web server allow arbitrary commands to be executed as "nobody." The result is that these SSIs can kill Apache's child processes, which also run as "nobody." Running the `kill` command in a loop results in a local denial-of-service attack against the Web server.

Summary Sheet—Running with Elevated Privilege

Problem:
Applications that operate at multiple levels of operating system privilege are prone to privilege elevation vulnerabilities. An attacker manipulates functionality within the application or causes a buffer overflow to force an elevation of privilege within the system.

Potential Impact:
An attacker might, in the worst case, gain complete administrative control of a system.

Habitat:
Applications that run in multi-user environments or need administrative privileges to accomplish a function are particularly prone to this vulnerability.

Tools You Need to Find It:
Ordinary OS tools such as `find` and `ls` can be used to identify potentially vulnerable applications.

How to Look for It:

The best way to identify these vulnerabilities is to review the source code surrounding the privilege escalation need.

Symptoms of Failure:

The application does "too much" in a privileged mode. The application fails to adequately constrain inputs. The application allows for programmatic or shell access in privileged mode.

Famous Failures/Exploits:

- `dtappgather`—A vulnerability in CDE (Solaris, AIX, HPUX) that allowed an arbitrary file to be copied.
- `uucp`—A buffer overflow permitted local execution of arbitrary code as a privileged user.
- `sendmail`—The "mother" of all buggy suid programs, `sendmail` has had several exploitable buffer overflows.

REFERENCES

[Bell73] Bell, D.E. and LaPadula, L.J., "Secure computer systems: Mathematical foundations and model." Technical Report M74 244, The MITRE Corp., 1973.

5 Permitting Default or Weak Passwords

In This Chapter

- Finding Default and Weak Passwords
- Fixing This Vulnerability
- References

Weak passwords have been called the biggest problem with information security. As passwords become more and more pervasive, users are compelled to remember more and more of them, meaning they are either choosing very simple ones, or using the same password over and over. Danger lies in either of these scenarios—simple passwords based on usernames, dictionary words, or four-digit numbers are very simple for an automated password cracker to discover; using a strong password that is the same on numerous systems dramatically increases the risk of all systems being compromised if that password is discovered. You might ask, "Why is password safety a developer concern and not a user concern?" The answer is that an attacker could use your application's lax security to recover passwords and attack other applications with them, or spread a virus or worm based on your application.

Passwords are common enough to be almost forgotten about. Whether it's a secret handshake, an ATM pin, or a bicycle lock combination, shared secrets are, perhaps with the exception of mechanical keys, the most common way of authenticating a person or thing to permit or deny access. When we think of computer usernames and passwords, we usually think of the logon authentication feature provided by the operating system—Unix's trademark `login:` prompt or Windows' login dialog. However, many applications besides the operating system offer authentication. Database applications, enterprise resource planning (ERP) systems, Web sites, and financial applications are all examples of applications that "roll their own" authentication; that is, they implement authentication within their application separate from the operating system. The selection of passwords is typically left to the user; this creates trust between the user and provider of the software because as long as the system permits arbitrary passwords, you have a lessened risk that someone with knowledge of the authentication algorithm could circumvent it. However, it is generally not wise to permit *all possible* passwords. Some are, for mathematical reasons, better choices than others.

Consider the Unix password scheme. Traditionally, a Unix password is between three and eight characters, and the permitted characters are printing characters between 21 and 7E hex. This creates a theoretical password space of 94^8 possible passwords. However, an early study by the creators of Unix [Morris79] discovered that a dictionary attack (trying all dictionary words as password candidates) was able to discover a third of their 3,289 sample passwords, and that an algorithm that tried one-, two-, three-, and four-letter alphanumeric combinations and five- and six-letter combinations, in conjunction with a dictionary attack, was able to obtain 86 percent of the sample passwords. So in practice, of the 6,095,689,385,410,816 possible passwords, the most common 322,671,313 combinations make up 86 percent of actual passwords. That's a 10^8 reduction in the password space.

TESTING THE SECURITY OF YOUR PASSWORD

Use this table to determine how many tries (in the worst case) it would take a brute force algorithm and dictionary attack to determine your password:

One-, two-, three-, or four-letter English word	2,263 tries
Five- or six-letter English word	8,226 tries
One, two, three, or four random alphanumeric characters	1,874,161 tries
Five or six random letters	320,797,152 tries
Six to eight random alphanumeric and non-alphanumeric characters	23,811,286,661,761 tries

Another problem exists when application developers ship their product with a vendor-supplied default password or no password. Many users do not change the default password. This happens regardless of how security conscious the intended users are, and how large and bold the warnings are in the manual. Some users don't trust authentication, especially during the install process. It becomes one more thing that can "go wrong" with the install. Shipping your product with an empty or default password ensures that the default password bubbles to the top of the list in password-cracking software dictionaries. These lists are updated frequently on the Internet and in hacker magazines like *Phrack* and *2600*.

Numerous vulnerabilities and incidents have resulted from the tendency of administrators to leave default passwords in place. Products ranging from the SAP enterprise resource planning software to consumer DSL routers have had this vulnerability reported in them. In the mid-1990s, a common hacker technique was to modify access control rules and routes in the networking hardware of their victims; many networking hardware vendors shipped their products with the default password "public." The SQLSnake vulnerability, which caused major havoc for corporations and backbone providers, took advantage of the default password setting shipped with earlier versions of Microsoft SQL Server. All of these vulnerabilities could have been prevented if the user had only been *required* to set a password to make the software operational. (The current version of Microsoft SQL Server *does* force users to set an "sa" password on install.)

FINDING DEFAULT AND WEAK PASSWORDS

Password crackers have some limitations. Most are advertised as a way to recover lost or otherwise unrecoverable passwords, though in order to use them for this purpose you must have some direct access to the password database. In short, you must already be "in"to use one of these. Another legitimate use of password crackers is to ensure that an operating system shipped as part of an integrated application did not contain any back door users or empty passwords.

Under Windows, one of the best programs for password cracking is the commercial application LC4 from @stake, Inc. (available at *www.atstake.com*). LC4 can automatically recover passwords from four sources: the local machine's password database, a remote machine on a domain to which you have administrator access, a recovery disk made when the operating system was installed, or passwords sniffed off the network. The last feature (passwords sniffed off the network) seems to be at odds with any legitimate use of LC4.

You can also select the strength of the audit, varying from a simple dictionary attack to an intensive combinatorial one. Selecting one of these retrieves the password database and begins trying candidates. Figure 5.1 illustrates the output of LC4.

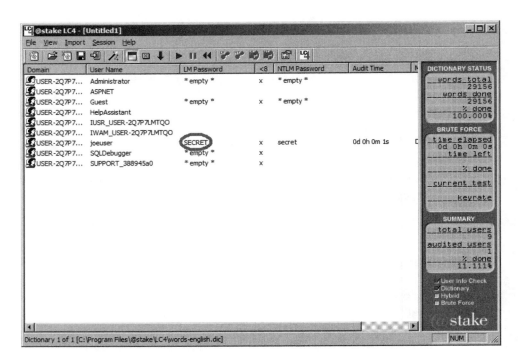

FIGURE 5.1 LC4 output.

On Linux, you can use the John the Ripper password auditing tool. John can crack Unix, DOS, Windows NT, and Windows 95 passwords and operates from the command line. You need to have access to, and supply to John, the password file you want to crack. A sample John session looks like this:

```
./john -users:simon /etc/shadow
Loaded 1 password (FreeBSD MD5 [32/32])
says        (simon)
guesses: 1 time: 0:00:08:16 (3) c/s: 3925 trying:
says
```

In this case, we "audited" the password for the user named simon. After 8 minutes and 16 seconds, John correctly guessed our four-letter password says.

Building a Password Cracker

ON THE CD

It is relatively easy to build your own password cracker. We have built one for the companion CD-ROM included with this book, in the Source Examples\Chapter 5 directory. Our tool performs a simple dictionary attack against Unix and Windows passwords. We make no claim that it is optimal; several improvements could be made, including precomputation of hashes, trying variations on words and names (such as "scott1" for "scott"), etc. Our purpose is to demonstrate the techniques for testing the validity of passwords and the functions used within operating systems to do this.

As we mentioned previously, one technique operating systems use to prevent brute force password attacks is to make the password database unavailable to an ordinary user. This was not true on older variations of Unix and continues to be true in some instances. Because the password database (/etc/shadow on Linux) is owned by root and is not readable, you can use our technique only on systems on which you already have administrator access. To use it on other kinds of systems, you must first gain root privileges. Figure 5.2 shows a typical Linux shadow file.

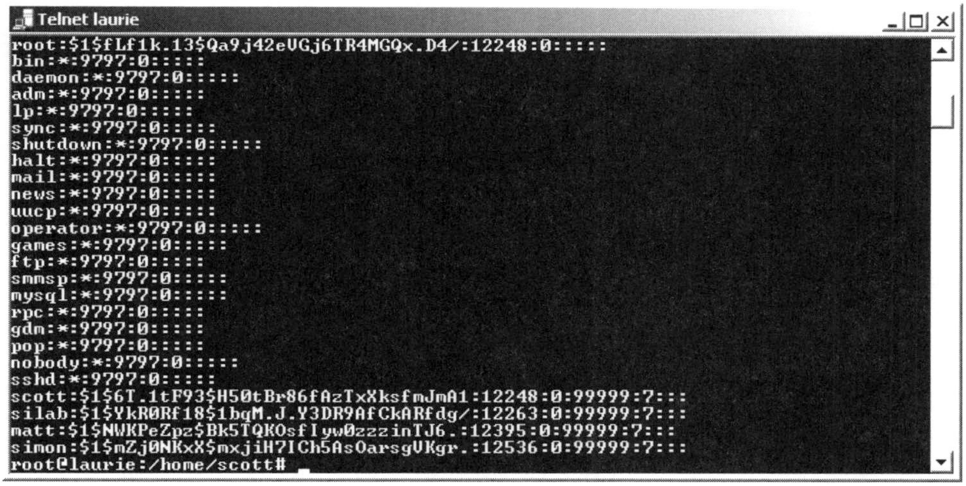

FIGURE 5.2 A typical Linux shadow file.

Windows passwords are located in the SAM database, which can be read via a mounted Windows share. Phil Staubs' pwdump3 utility (obtainable from *www.openwall.com/passwords/nt.shtml*) is capable of doing this.

Once the password database is obtained, test passwords can be run against it. This need not be done on the same machine that the password database resided on, because our routines do not attempt to actually authenticate on the machine. Figure 5.3 shows the output of pwdump3.

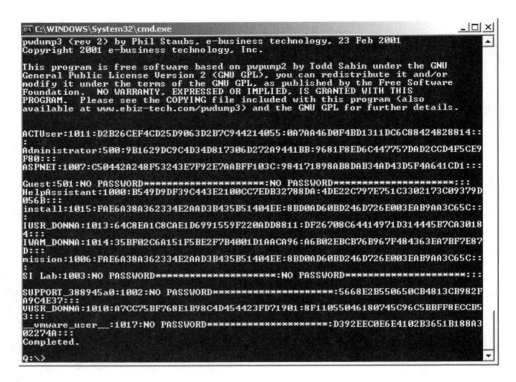

FIGURE 5.3 The output of pwdump3.

Using a Dictionary Helper

To generate candidate hashes, it is necessary to have a dictionary of common passwords. We obtained a number of word lists including the most common English words, most common proper names, and other popular password choices, including names of pets and animals, names of sports teams, etc. We combined these word lists into a single file, all.txt, and eliminated duplicates. We wrote a small dictionary helper class that allows us to interact with the dictionary.

```
#define MAX_WORD_LEN 80

class pwDictionary {

public:

    FILE* fv;
```

```
        pwDictionary();
        void setFile(char* fname);
        int nextWord(char* word);

};

void pwDictionary::pwDictionary()
{
    fv = 0;
}

void pwDictionary::setFile(char* fname)
{
    fv = fopen(fname,"r");
}

int pwDictionary::nextWord(char* word)
{
    if(!fv) return -1;

    char theWord[MAX_WORD_LEN];

    if(feof(fv))
    {   fclose(fv);
        return -1;
    }

    fgets(theWord, MAX_WORD_LEN-1, fv);

    theWord[MAX_WORD_LEN-1]=0;

    if(strlen(theWord) >= MAX_WORD_LEN-2)
        return -1;

    if(feof(fv))
    {
        fclose(fv);
        return -1;
    }

    if(strchr(theWord,'\n'))
        *(strchr(theWord,'\n')) = 0;
```

```
        strcpy(word,theWord);

        return strlen(word);
    }
```

The class has two methods: `setFile()` and `nextWord()`. (The constructor `pwDictionary()` simply sets the file pointer to zero.) `setFile()` opens a file given by the filename. If this call fails, subsequent calls to `nextWord()` also fail.

`nextWord()` obtains the next word in the file. Words are by themselves on a single line. Some error handling is required in case of an inadvertent blank line or missing carriage return within the file.

The basic idea is that the password cracking routine can call `nextWord()` to obtain the next candidate password to try. If need be, it can load multiple dictionaries by calling `setFile()` multiple times.

Writing the Main Crack Routine

Our main routine, `crack()`, is written to handle both Unix and Windows style hashes. This is because the behavior of the cracking routine is the same regardless of the platform; only the mechanism for computing the candidate hash is different.

`crack()` takes three arguments. The first is the key (encrypted hash of the real password) that we are attempting to crack. The second is the name of a dictionary file to be passed to `pwDictionary::setFile()`. Finally, it takes a predefined constant, either `CRACK_UNIX_CRYPT` or `CRACK_W_MD4`, which determines whether the Unix style (DES/MD5) or Windows (MD4) style hashing algorithm is used.

```
void crack(char* key, char* dictionary, int alg)
{

    pwDictionary dict;
    dict.setFile(dictionary);

    int res = 0;
    char candidate[80];
    bool b;
    int ct=0;

    while(res!=-1)
    {

        res = dict.nextWord(candidate);
        if(res==-1) break;
```

```
        if(alg==CRACK_W_MD4)
            b = try_one_windows_password((unsigned char*) candidate,
(unsigned char*) key);
        else if(alg==CRACK_UNIX_CRYPT)
            b = try_one_unix_password((unsigned char*) candidate,
(unsigned char*) key);
        if(b)
        {
            printf("password is `%s'\n",candidate);
            return;
        }
        ct++;


    }
    printf("failed after %d guesses\n",ct);

}
```

`crack()` loops through each word in the dictionary, calling either `try_one_unix_password` or `try_one_windows_password`, which both return true if the hashes match.

The Windows hashing function is easier to write, because it is just an MD4 hash. We obtained the MD4 hashing function included in the open source Samba package for Linux (*www.samba.org*), which contains the function E_md4hash. Writing your own MD4 hashing function is an exercise in mathematics, not security, and it is better to obtain one of the freely available ones.

```
bool try_one_windows_password(unsigned char* passwd, unsigned char* key)
{
    unsigned char p16[16];
    char temp[3];   // for the two digits of hex, plus a trailing \0
    char res[64];

    E_md4hash(passwd, p16);

    strcpy(res,"");

    for(int i=0;i<=15;i++)
    {
        sprintf(temp,"%02X",p16[i]);
        strcat(res,temp);
    }
```

```
        return !strcmp(res, (char*) key);

}
```

Some formatting is needed to convert the results of E_md4hash() into the human readable format we are using. To do this, we effectively "print" each byte using sprintf and compare the results.

Building the Unix function is more difficult, because modern Unix systems accommodate both traditional DES and modern MD5 hashes. The first two characters of the hash are a two-character salt, used to perturb the DES algorithm one of several different ways. The salt "$1" is a special code that indicates the password was hashed using MD5 rather than DES. Depending on this value, we call either crypt() or md5_crypt().

```
bool try_one_unix_password(unsigned char* passwd, unsigned char* key)
{
    char salt[14];

    salt[0]=key[0];
    salt[1]=key[1];
    salt[2]=0;

    if(strcmp(salt,"$1")) // DES; easy case!
    {
        char* hash = crypt( (char*) passwd, (char*) salt);

        return !(strcmp( (char*) hash, (char*) key));
    }
    else // MD5; kind of hard!
    {
        strncpy(salt,(char*)key,14);
        char* p = strchr(&salt[4],'$');
        if(p) *(p+1) = 0;

        char* hash = md5_crypt( (char*) passwd, (char*) salt );
        return !(strcmp( (char*) hash, (char*) key));


    }
```

```
    return 0;

}
```

The salt for MD5 can be up to 14 characters long and is contained at the beginning of the key following the "$1". It begins and ends with a $. This salt must be separated from the key proper before calling md5_crypt().

Putting It Together

To crack simon's password, we call

```
crack("$1$mZjONKxX$mxjiH7ICh5AsOarsgVKgr.", "all.txt",
  CRACK_UNIX_CRYPT);
```

One of two results occurs. If the password is obtainable, the program outputs password is: followed by the correct password. Otherwise, it outputs failed after <n> guesses where <n> is the number of words in the dictionary.

In fact, it correctly guesses simon's password to be says (as John did in the previous example) and does so in 65 seconds on our modest 1 GHz machine. As we mentioned previously, significant improvements could be made including prehashing and indexing of candidates, though LC4 performs adequately and is recommended should the need for a password cracker arise.

FIXING THIS VULNERABILITY

Applying a few simple rules can prevent your application from becoming vulnerable to a password attack.

- Don't ship with a default password or blank password. *Require* a password be entered prior to operation, preferably at install, and ensure that the software does not allow the administrator to set a blank password.
- Limit direct access to the password database and lock the system for a period of time after an unauthorized access. This period of time does not need to be very long to inhibit automatic password cracking tools.
- Enforce strong password selection on the user. Don't permit the user to select an arbitrary password, or if you must, require that the administrator change the default security settings for the application before you accept a weak password.
- Consider automatically generating a password and supplying it to the user, or requiring two credentials, such as an automatically generated password and

user selected PIN number to gain access. To succeed in gaining access, an attacker would have to guess *both* of these correctly.
- If your system supports very long (e.g., up to 128) characters, encourage users to choose a short phrase or sentence rather than a single word password.

GENERATING PASSWORDS A USER CAN REMEMBER

This routine generates a password composed of two four- or five-letter words, separated by a non-alphanumeric character. The standard Linux dictionary has 5,049 four- and five-letter words, so it has a password space of more than 254,000,000 combinations. The companion CD-ROM contains the code to this routine along with the passgen.in file containing the dictionary words.

```c
/* program to generate secure, easy to use passwords */

#include <stdio.h>
#include <sys/time.h>

/* passgen.in has 5049 entries */
#define wordct 5049

char words[wordct][7];
char* separator = "+-=*$#%@&!";

void main()
{
  FILE* wordfile;
    int i, word1, word2, sep;

  wordfile = fopen("passgen.in","r");
    if(!wordfile) exit(-1);

  for(i=0;i<=wordct-1;i++)
  {
    fscanf(wordfile,"%s\n",words[i]);
  }

  fclose(wordfile);

  srand(time(NULL));
```

```
    word1 = rand() % wordct;
    word2 = rand() % wordct;

    sep = rand() % strlen(separator);

    printf("Your system-generated password is ");
    printf("%s", words[word1]);
    printf("%c", separator[sep]);
    printf("%s", words[word2]);
    printf(".\n");

}
```

Summary Sheet–Permitting Default and Weak Passwords

Problem:

Users are overwhelmed with the number of passwords they must remember on a daily basis. Whenever possible, users select short, easy to remember passwords. This tendency makes it easy for an attacker to apply a brute force attack that recovers the password.

Potential Impact:

An attacker could gain access to confidential data or execute commands he is not authorized to access.

Habitat:

Any application that uses a password to control access to or otherwise protect data and application functionality could potentially be at risk for a brute force password attack.

Tools You Need to Find It:

A password cracking tool and access to the password database. A memory search tool, such as the search feature of OllyDbg, is useful for finding passwords in memory.

How to Look for It:

Look for passwords that are too short, contain only letters or numbers, or are composed of dictionary words.

Symptoms of Failure:

Password cracking software can recover the password in a relatively short period of time. The period of time depends on the sensitivity of the information and the willingness of the attacker to apply resources to the problem.

Famous Failures/Exploits:

The Morris Internet worm, the Internet's first major security incident, worked because some Unix machines had empty root passwords. Don Sealy wrote an analysis of the worm, which is preserved at SecurityDigest. [Seely89] SecurityFocus describes the cause and effects of the SQLSnake worm. (*www.securityfocus.com/news/429*)

REFERENCES

[Morris79] Morris, Robert and Thompson, Ken. "Password Security: A Case History." *Communications of the ACM*. Vol. 22, Number 11. 1979: pp. 594–597.

[Seely89] Seely, Donn. "A Tour of the Internet Worm." Available online at *http://securitydigest.org/phage/resource/seely.pdf*.

6 Shells, Scripts, and Macros

In This Chapter

- Description
- Fixing This Problem
- References

For most of the history of computing, including the early days of personal computers, the ordinary user was a programmer. Whether it was preparing punch cards for use on an IBM mainframe or tinkering with BASIC on the Apple II and Commodore 64, programming was at the core of computing for the user as well as the software maker.

The ability to program a computer in its native hardware language or a high-level language is not required of today's user. However, even today's sophisticated applications don't necessarily offer all of the features that a user requires, or the ability to do repetitive tasks efficiently. Because of this, many applications provide scripting or macro languages that can be used to program them beyond the functionality available from the pull-down menus and other user interface functions. Some of these languages are quite simple and are confined to record and playback of user interface (UI) sequences or application functions. Others, such as the Microsoft Office scripting environment, are really high-level languages in their own right, adapted to suit application programming. Often times these applications embed macros or scripts within the application's document or data file. The result is a document that contains both data and executable logic.

Convenient as it is to the user, the problem with this programmatic access to applications is that an attacker can use it as well. Consider the example of a word processor that can embed Visual Basic code within the document and execute code when the document is opened. A user does not typically think of word-processing documents as malicious programs. The result is that a user might unsuspectingly open the document and have arbitrary commands executed on his machine. This is the root cause of most macro viruses, and while the application programmer might blame the user for foolishly opening the document, these viruses have been one of the most costly kinds of security problems in recent history.

In this chapter, we discuss programs that provide command shell access, macros, or scripting capability. As we'll see, the macro virus represents only one prospective security problem arising from this programmatic access.

DESCRIPTION

Shell, script, and macro vulnerabilities have been among the most destructive and expensive security flaws to date. Consider the following examples:

- The Melissa virus, a Visual Basic macro virus, is estimated to have cost $80 million to repair and clean up. [Festa00]
- "I Love You," a virus similar to Melissa that reached a much larger number of users, might have ultimately cost businesses $2.6 billion in lost productivity, repair cost, and lost business. [Festa00]
- Browser-based "phishing" attacks, facilitated by scripting bugs in Web browsers like Internet Explorer, were responsible for a significant portion of the $2 billion consumers lost to electronic theft and fraud in 2002–2003. [Sullivan04]

These vulnerabilities have numerous variants and continue to circulate, meaning the real costs will never be known. The problems are not confined to Microsoft, either. Macintosh and Linux users, and users of alternate browsers such as Firefox, have also been subjected to phishing attacks.

An attacker needs two things to make a scripting attack successful. First, the intended target must have the vulnerable application, and the attacker must be able to get the script to execute within the application. This might be through embedding it in an otherwise innocuous file, tricking the user into executing it by using deceptive naming or social engineering, or exploiting a mechanism that forces the application to execute the script automatically. One virus copied itself over every image file on the machine and renamed that file `filename.JPG.vbs`, knowing that by default Windows Explorer hides file extensions for known types. In this way, it was able to propagate every time a user tried to open the image. Other attacks have taken advantage of an application's event model, hijacking the `OnLoad` or `OnExit` events to ensure that the macro is automatically executed if the file is opened (or closed).

Second, the script provider (the application in concert with its macro language) needs to provide functionality of interest to the attacker. A macro language for a presentation tool that allowed the user to script only the timing between slides, for example, would not generally be useful to a virus writer. An application that had a Visual Basic (or Perl, Scheme, REXX, etc.) back end or allowed a user to execute shell commands would be much more dangerous. Even without a rich command set, it is still possible for an attacker to exploit the system. The script might be able to gather seemingly innocuous information like the names of recent documents or the version of the software, which can be combined with another attack. (Sometimes this data, especially if it contains financial information, is what the attacker is looking for.) Depending on how the script language is implemented, the attacker might be able to intentionally code an exploitable buffer overrun into his code, or import an OS function that has one in it. A macro that is executed by a program in a shared environment (such as a Web server) might be able to crash the application for all its users. Finally, sometimes macros have access to test APIs or functions that cannot be reached through the user interface. This is because QA testers at the company that wrote the software frequently use the application's own scripting language to test it. These features might be able to bypass some security within the application.

Shell, script, and macro vulnerabilities occur in many forms; document-based macro viruses are only one example. Each kind of macro implementation lends itself to its own set of potential vulnerabilities.

Embedded Script Languages and Command Interpreters

These problems result from applications that link to a general-purpose programming language or provide limited shell script access in order to accomplish a function. Examples include Visual Basic for Applications embedded within Microsoft Office or applications that support scripting in TCL or Python. To provide maximum flexibility, or because they are relying on the script language as an add-on library, these applications don't limit the functionality of the interpreter enough to prevent malicious code from running.

Document Markup

Sometimes macro commands can be embedded within a document. When the user opens the document, the commands are automatically executed. *Server Side Includes* (SSIs) within Web pages are commonly implemented this way. A user embeds a command within the HTML document in what looks like an HTML comment:

```
<!-#command tag1="value1" tag2="value2" ->
```

The `exec` command allows execution of an arbitrary shell command in the command interpreter.

```
<!-#exec cmd="ls" ->
```

would execute the `ls` command and include the results within the Web page. While SSIs are often thought of as safe, they can, for example, read and write files belonging to the user "nobody." They can also be used as a vector for deploying a known local privilege escalation exploit by a user who does not have shell access, or for creating a denial-of-service attack by running the system out of resources.

Earlier versions of the Acrobat Distiller program could execute shell commands embedded within a PostScript document. PostScript is a language spoken by many printers and is the default printing format in the Unix and Macintosh communities. Academic publications and business forms are also frequently laid out in PostScript. Some of the same features that made PostScript appealing as a printer control language made it undesirable from a security point of view. For example, a PostScript file can contain a command such as:

```
%!
(myfile.txt) deletefile
showpage
%%EOF
```

If an attacker could get a user to open a file with this command embedded in it, the effect would be to delete the file named `myfile.txt`. Because PostScript can also create files, the attacker could place a Trojan executable in the Windows startup path, or overwrite a security-relevant file.

JavaScript

JavaScript, the embedded script language that, along with its feature-for-feature cousin VBScript, can be used to script the Web browser has been plagued by security issues since it was introduced by Netscape in 1995. JavaScript is unique among script hosts in that it is specifically designed to be used by "someone else" to manipulate an application on your computer. Because of this, JavaScript's designers had to make tradeoffs as to what should be allowed and what should be prohibited to the script host. Previous versions could read and write files, obtain information from the document object model, and repost that information to another site, and overload events to prevent windows from closing or being resized.

JavaScript manipulation represents a danger to any application that is hosted in the browser. This is especially true as browser-based applications running purely on the client have emerged, notably HTML help. Additionally, the new version of the Windows client, "Avalon," to be released with the "Longhorn" version of Windows, will blur the line between conventional and browser-hosted applications [Arar03]. Chapter 18 describes one way that a malicious user can force JavaScript to execute within someone else's application.

Safe for Scripting ActiveX Controls

As their name implies, these are ActiveX controls that are marked "safe" to be executed in the browser environment, where they can be manipulated by JavaScript. The "safe for scripting" attribute is intended to be used by controls that are primarily designed to be run in the browser environment: the Macromedia Flash plug-in is a good example of one. However, due to the flexible nature of ActiveX, any script marked safe can be invoked, regardless of the functionality contained in that control. Nothing within the browser prevents a control marked "safe for scripting" from manipulating the local filesystem or exercising any part of the Win32 API. The CERT and CVE databases are full of ActiveX controls that are marked safe that shouldn't have been.

Database Stored Procedures

Some database packages contain stored procedures that can be used to perform tasks beyond those supported by the standard SQL query language, notably Microsoft SQL Server's `xp_cmdshell()` stored procedure, which allows for execution of arbitrary DOS shell commands from within the database application. Applications that use a database need to be careful when processing user input and forming commands to ensure that calls to stored procedures cannot be inadvertently made.

Macro Expansion in Logs and Messages

Some logging components of applications allow for macro processing. For example, an application that wanted to log an error message with the current user's username might pass just %u to the logging application; the logging application would then look up the username and expand it into the log file.

While these applications typically do not provide a very rich macro language, they might be susceptible to buffer overflow attacks if they allow expansion of arbitrary length strings. The syslog-ng product from Balabit had this problem. It allowed arbitrarily long constant character strings to be expanded in the syslog; an exploitable buffer overflow resulted from this (*www.securityfocus.com/bid/5934*).

FIXING THIS PROBLEM

Each kind of script and macro vulnerability calls for its own solution. However, a few general-purpose solutions fix this problem across different kinds of vulnerabilities and applications:

- Some application scripting languages have features that reduce the risk of script injection. For example, Perl 5 has the "taint" option, -T, that treats all user-supplied input as suspicious. When Perl is invoked with this option, it prevents a number of "risky" scenarios such as invoking a shell with user data as a parameter. Another Perl feature, "use strict," forces the developer to declare all variables before using them, which prevents misinterpretation of the source of potentially malicious data.
- If your application must support programmatic access, turn it off by default. Most users won't require macro functionality because it requires some programming knowledge to use in most cases. By turning macro functionality off by default, you are significantly reducing the chances of any given installation of the application being vulnerable to attack. Alert users to the security risks of macros both at install time and when the user selects/unselects macro functionality. Better yet, consider whether your application needs macro support at all. Why do you support macros? Because the *programmers* wanted part of the application to be programmable, even if the users don't. (This is often the main reason why macro support is found within applications.) Can you accomplish the same goal another way (such as record/playback within the UI or extensions to the file format)?
- When in doubt, use a visual prompt to warn the user of potential malicious action. Sometimes this is all that is needed. However, this solution is not perfect. A user can be "socially engineered" to ignore these messages either by too-frequent prompting or when the user has sufficient desire to continue (offer of free money, etc.).

- The prevalence of viruses like "I Love You" has shown that users ignore warnings that macros might be insecure. One way to limit the *effect* of these viruses is to prevent macro functionality from accessing the network. For example, a word processing program with built-in e-mail capability could disable the e-mail functionality while the macro interpreter is running. Thus, regardless of how much damage the macro can cause to the individual's machine, it cannot spread itself. This takes away the user's ability to script the sending of e-mail messages, but isn't that the point? This solution does have some limitations. For example, an attacker could manipulate the timing of the script by, say, scheduling an operating system event to send the e-mail later.
- Ensure only "white" data is passed from the network to a macro processing application. Data should be free of embedded commands and excessively long strings. Chapter 18 describes ways to sanitize HTML documents to remove script and macro code.
- Use a "sandbox" to prevent malicious behavior. Do not allow scripts or macros to have complete access to the filesystem. You can use a "chroot jail" in Unix to prevent access to the regular filesystem; in Windows this would have to be built into the application via selective filtering of filenames. Assume any script language that can launch a shell command or create an arbitrary file can be used to exploit a system.
- Avoid implementing scripting through a "full-fledged" programming language like Visual Basic, Perl, or Tcl. Unless you have a way to lock these down so they can't perform most OS functions, you're giving the user way too much power here.

Summary Sheet—Shells, Scripts, and Macros

Problem:

Code and data are often packaged together. This code can take the form of scripts or macros that are executed by the application. Because some users might open a file containing macros or scripts despite all warnings, measures have to be taken to ensure that no malicious side effects can result from macros.

Potential Impact:

An attacker could execute arbitrary commands on the system.

Habitat:

Applications that support macros, programmable components, extensibility, scripting languages—in short, any program that interprets data as though it were code.

Tools You Need to Find It:

Macros and scripts are vulnerabilities at the design level. Design inspection is the best way to find macro vulnerabilities.

How to Look for It:

Examine the file format and menus of an application for evidence of script or macro capability. Examine the log files an application produces. Do they contain user-supplied data? Does the documentation call out macros as a feature of the application? Watch for applications that install known scripting hosts such as Visual Basic for Applications, Perl, or Tcl.

Symptoms of Failure:

Try finding a way to write a file to an arbitrary location or to execute a shell command. These are good indicators that the system is vulnerable.

Famous Failures/Exploits:

The Melissa virus and I Love You virus were the result of malicious scripts in the Microsoft Office package.

REFERENCES

[Arar03] Arar, Yardena. "Sneak a Peek at the Next Windows." *PCWorld.com.* Available online at *www.pcworld.com/news/article/0,aid,113129,00.asp*, October 27, 2003.

[Festa00] Festa, Paul and Wilcox, Joe. "Experts estimate damages in the billions for bug." *Cnet news.com.* Available online at *http://news.com.com/2100-1001-240112.html?legacy=cnet*, May 5, 2000.

[Sullivan04] Sullivan, Bob. "Survey: 2 million bank accounts robbed." *MSNBC News.* Available online at *www.msnbc.msn.com/id/5184077/*, June 14, 2004.

7 Dynamic Linking and Loading

In This Chapter

- Finding This Vulnerability
- Fixing This Vulnerability
- References

When we develop software, we often assume that the same functions we intended to call are actually called by our program. That is to say, if we choose to call `printf` with a certain set of arguments, we expect the real, standard C library version of `printf` is called. If we include all of the functions that our application uses within the executable—a method known as *static linking*—this is indeed the case, and the result is a large, monolithic executable. We have some advantages in this approach; namely, we can deploy and install applications easily, by just moving a single executable. More typically, however, applications ship with multiple libraries, and they also rely on libraries from the operating system and third parties. When the application is invoked, it loads some of these libraries immediately and loads others when needed. This process is referred to *dynamic linking and loading*.

We certainly find advantages to this model, too. The application is highly modular, meaning that updates and patches can be significantly easier to deploy. Also, the application can take advantage of updates to the contents of operating system functions and third-party libraries.

With this modularity, however, comes risk—a risk that a user can replace a legitimate library with one modified by an attacker. Sometimes an attacker can escalate his privilege by forcing a privileged process to execute attacker commands by fooling the process into loading the attacker library instead of the intended library.

This attack is usually facilitated by *library search order*. When a library is loaded by name—such as a call to load the library kernel32.dll in Windows—the application follows a set search sequence when it looks for that file. In many incarnations of Windows, for example, the OS searches the following locations [Microsoft03] (in order) for the library:

1. The directory from which the application loaded
2. The current directory
3. The system directory
4. The 16-bit system directory
5. The Windows directory
6. The directories that are listed in the PATH environment variable

Windows Server 2003, Windows XP SP1: The default value of `HKLM\System\CurrentControlSet\Control\Session Manager\SafeDllSearchMode` *is 1 (current directory is searched after the System and Windows directories). Windows XP: If* `HKLM\System\CurrentControlSet\Control\Session Manager\SafeDllSearchMode` *is 1, the current directory is searched after the System and Windows directories, but before the directories in the PATH environment variable. The default value is 1 (current directory is searched after the System and Windows directories). Note that this value is cached on a per-process basis at load time.*

An attacker could create a library that the application finds and loads instead of the intended library, possibly forcing a privileged process to execute instructions placed in an imposter library by a less privileged user. In addition to executing arbitrary instructions, the impostor library might allow an attacker to overwrite methods in the application under test.

While this is obviously a concern for processes that run with higher privileges than the user, several other risks exist. Consider, for example, the risk of having a key logger installed on a system that is publicly used. Methods for checking for keystroke loggers exist, and most detection methods assume that the logger is running as a separate application that is visible through the process list. However, imagine the functionality to manage Internet forms for a particular Internet browser were implemented in a DLL. If an attacker were to Trojan that library, password data from all forms could be siphoned off and sent to an attacker's Web site with no detectable key logger running.

Applications that implement anti-debugging are also at great risk of dynamic linking and loading attacks. Consider applications that enforce Digital Rights Management (DRM). Many of these applications include code to search for the presence of tools that are trying to inspect the running process like debuggers. If care is not taken, an attacker can replace one of the libraries loaded by this application with ones that inspect memory and save its contents off without the need of a debugger.

Dynamic linking and loading also raises the need to ensure that libraries actually load. Consider, for example, the "Content Advisor" feature in Microsoft's Internet Explorer® Web browser. It allows a user to control the type of sites that others who use the machine have access to on the Internet by password protecting individual sites, categories of sites, or unrated sites, as shown in Figure 7.1. The security risk here is allowing access to a prohibited site.

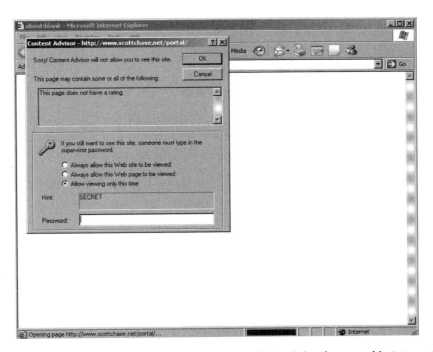

FIGURE 7.1 Security in terms of access restriction is implemented in Internet Explorer.

Internet Explorer loads several libraries at runtime. Figure 7.2 shows an observation tool called Holodeck that lists the libraries loaded by Internet Explorer at runtime. Several Windows operating system libraries are loaded, too, including Kernel32.dll, User32.dll, and UxTheme.dll. Among those libraries is the file MSRating.dll. Holodeck is available for trial download from Security Innovation® at *www.securityinnovation.com/holodeck/*.

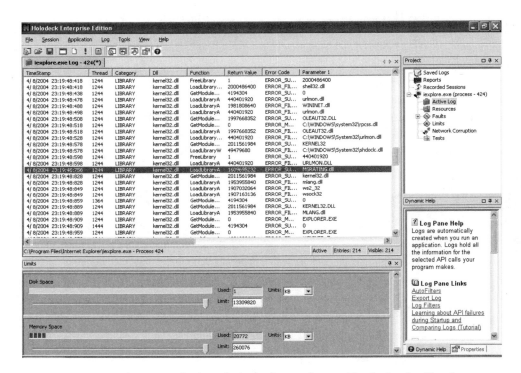

FIGURE 7.2 Security Innovation's Holodeck observation tool looks for the libraries loaded by Internet Explorer.

Once we notice that an external dependency to determine whether or not to grant access to a site exists, this signals a potential point of failure that should be investigated. We can use Holodeck to block access to the MSRating.dll by forcing the LoadLibraryA and LoadLibraryW system calls made by Internet Explorer to fail when they attempt to load Msrating.dll (Figure 7.3).

In this environment we then navigate to a blocked Web site. Internet Explorer opens the page without prompting for a password. Figure 7.4 shows that if we then go into the Internet Options menu, we are now unable to alter the content settings and the relevant buttons are not selectable; security measures have been bypassed and the user is now free to explore the Internet unchecked.

Dynamic Linking and Loading

FIGURE 7.3 We can block Internet Explorer's access to the MSRating.dll library using Holodeck's Scheduled Test feature.

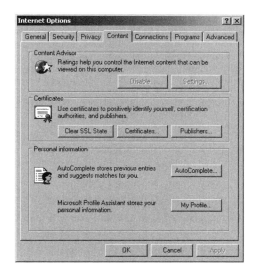

FIGURE 7.4 Blocking the return value of MSRating.dll completely subverts the content rating feature of Internet Explorer.

In this case we were able to bypass security controls by simply denying an application the capability to load a library dynamically. A more common attack vector is to replace the target library with one of an attacker's choosing and force the application to execute some of the instructions in the impostor DLL. If such a vulnerability exists, the likely consequence is being able to escalate privilege over an application or the system.

FINDING THIS VULNERABILITY

You have several techniques for finding dynamic linking and loading issues through testing. A first approach is to examine how an application handles a failure to load particular libraries. From a security standpoint, our goal is to deprive the application of either security or validation functionality by depriving the application of external libraries that provide this service. One of the easiest ways to test this is to use Security Innovation's *Holodeck*, available at *www.securityinnovation.com/holodeck*. Holodeck allows a user to observe all of the library loads made by an application (as shown in Figure 7.2). Figure 7.3 illustrates the process of blocking a library load with Holodeck's Scheduled Test feature. If the application does not raise an error message or crash after the library load has been blocked, chances are it is not aware that it has been deprived of this functionality. Similar results can be obtained on other operating systems by simply deleting the library. Failures here are likely to take the form of some security or validation routine not being performed, so pay special attention to the security restrictions while the application is running normally and then verify that these restrictions still hold with blocked access to the library. A common failure is not encrypting data because an encryption library is unavailable. The result might be sensitive data being sent in plain text over the network or being stored unencrypted in files.

Another scenario that must be tested is attacking the application by forcing it to load an imposter library. This can be done by either leveraging the library search order or simply replacing the original library with a contrived one. This can be a powerful attack technique to escalate privilege. To be successful, the prototypes of the functions exported by the impostor library might need to match those of the original library. Also, some weak validation check of the library might be circumventable, such as storing a checksum of the library in a file or in the registry. Additionally, replacing system libraries can be protected by the Windows System File Protection, which silently restores any system files that have been marked as protected if a user or untrusted process attempts to overwrite them.

It is critical to test for this vulnerability in applications that have anti-debugging controls (common in applications that support Digital Rights Management).

Implementing anti-debugging in an application usually means that some sensitive data is passing through memory that an attacker should not be able to observe. If the application can be tricked into loading an impostor library, however, it is often the case that functions in the library can siphon off data in memory.

FIXING THIS VULNERABILITY

If an application must load libraries dynamically, several coding practices can reduce or eliminate the risk of library manipulation. One of the easiest is to check the return values of calls to the operating system to load a library. Applications that store validation code in a library might rely on the library performing some functions automatically on load (such as commands contained in DLLMain). In these cases, some problems can be avoided by checking the return value of the library load function to make sure it succeeded. This, of course, does nothing to protect against tampering with the DLL.

Rather than allowing the dynamic linker to resolve links to functions within a DLL, you can always link against a DLL explicitly. *Explicit* linking is when the program itself calls operating system functions to obtain the DLL and function handle. This prevents the problem of Trojan DLLs in the current or SystemRoot directory. However, it does not protect against one of those DLLs being replaced.

Another option is to perform a checksum, a hash, or other validation checks on the library. Before the library is loaded, the application can first validate that it has not been tampered with by comparing a value computed by analyzing the library (such as a hash) with a stored value. Care must be taken, however, to protect the "key" with which the library's hash or checksum is compared. In cases where the library must be loaded at invocation, the library can be validated (by comparing a checksum or hash) before its functionality is actually used. Chapter 12 shows examples of hashing functions that can be used to protect data. These same functions can be used to create a checksum of a DLL that can be checked when the program is started. However, this presents a different problem. Static verification of a checksum makes it significantly more difficult to patch a vulnerability (or any bug for that matter) in a DLL once that checksum is "set in stone." VeriSign® offers an Authenticode signing capability that overcomes this problem. Authenticode is described at *www.verisign.com/products-services/security-services/code signing/index.html*.

Explicit Linking and Loading of a DLL

It is relatively easy to link against a DLL explicitly. In Windows, the `LoadLibrary` and `GetProcAddress` functions are used to do this.

The first thing we need to do is declare a function pointer of the same type as the function we want to explicitly link. The prototype for the `strcpy` function in the standard C library is:

```
char *strcpy(char *dest, const char* src);
```

The declaration of a function pointer that matches this prototype would look like this:

```
char * (*myStrcpy) (char *, char *);
```

The `LoadLibrary` function is used to obtain a handle to the DLL we want to link against. Because `strcpy` is contained in `msvcrt.dll`, we call `LoadLibrary` with this argument:

```
HMODULE pMsvcrt;
char msvcrtName[256];
char* systemRoot;

systemRoot=getenv("SystemRoot");
strcpy(msvcrtName,systemRoot);
strcat(msvcrtName,"\\system32\\msvcrt.dll");

pMsvcrt = LoadLibrary(msvcrtName);
```

We use the value of the `SystemRoot` environment variable to find the true copy of `msvcrt.dll`. This is because `LoadLibrary` uses the same default search path when explicitly loading a DLL as the runtime linker does when it implicitly loads one.

Once we have obtained a handle to the DLL explicitly, we must retrieve a pointer to the function we want to call. `GetProcAddress` is used to do this. In our example,

```
myStrcpy = (char* (*)(char*,char*))
GetProcAddress(pMsvcrt,"strcpy");
```

would set the function pointer `myStrcpy` to the location of the real `strcpy` function. Our function pointer can now be called in the same manner as the real `strcpy`.

Summary Sheet—Dynamic Linking and Loading

Problem:

Libraries provide applications access to functionality not contained in the executable. Applications often ship with multiple libraries but applications also load operating system libraries as well and libraries from third-party components. Loading libraries dynamically creates the potential for an attacker to either deny the application access to the library or force the application to launch a potential malicious library in its place to gain control over the application's process.

Potential Impact:

Escalation of privilege, application functionality manipulation, and possible disclosure of sensitive information.

Habitat:

Desktop and server applications, that load libraries dynamically.

Tools You Need to Find It:

For Windows applications, Holodeck can be used to inspect the application's environment and library activity. For other platforms, debuggers can be used to list the modules loaded.

How to Look for It:

The first step is to identify which libraries are loaded by the application. This information can be retrieved using a debugger or a process inspection tool such as Holodeck. The next step is to see how the application responds when the application is deprived of these libraries and also test to see if the application can be forced into loading a different library than the one intended.

Symptoms of Failure:

When a library load is failed, application security failure takes the form of some security or validation function not being performed. When an application is forced to load an impostor library, and no overt error or crash indicates that the application is unaware that it has loaded an impostor library, this is a symptom of a dynamic linking and loading vulnerability.

Famous Failures/Exploits:

Bugtraq ID 72086: Executable path-searching vulnerability in Windows NT/2000. The default directory search order in Windows NT/2000 potentially allowed the local or remote execution of Trojan programs in other user accounts including Administrator (*www.securityfocus.com/archive/1/72086*).

REFERENCES

[Howard02] Howard, Michael and LeBlanc, David, *Writing Secure Code*, 2nd Ed. Microsoft Press, 2002.

[Microsoft03] "Load-Time Dynamic Linking." MSDN Online. Available online at *msdn.microsoft.com/library/default.asp?url=/library/en-us/dllproc/base/load_time_dynamic_linking.asp*. Accessed April 3, 2005.

Part III
Data Parsing

8 Buffer Overflow Vulnerabilities

In This Chapter

- Stack Overflows
- Exploiting Stack Overflows
- Heap Overflows
- Exploiting Buffer Overflows: Beyond the Stack
- Finding This Vulnerability
- Fixing This Vulnerability
- Endnotes
- References

Buffer overflow vulnerabilities account for a significant portion of all exploited security flaws in software. Many of the world's most devastating viruses and worms such as Code Red, Linux.Slapper.Worm, SQL.Slammer, and MS-Blaster have exploited buffer overflows in network enabled software or platforms to attack systems and spread. These flaws are among the most dangerous in software because many can be used (and have been) to completely take over a victim's machine by executing arbitrary instructions.

The consequences can be dire. Imagine the ability to steal all document files on a machine by sending a few network packets, or formatting a Web server by entering data through a Web form. In fact, frameworks such as the Metasploit Project (*www.metasploit.org*) exist to enable attackers (as well as security researchers) to automatically inject shellcodes that produce a desired behavior into newly discovered exploits. Figure 8.1 shows the Metasploit console.

FIGURE 8.1 Metasploit, a framework for automatically exploiting buffer overflow vulnerabilities.

Buffer overflows occur in software written in programming languages that do not strictly enforce bounds checking on arrays. The basic concept of a buffer overflow is that we provide an application with more data to be stored in a particular variable than the programmer set up space for. When this happens, it is likely that the application writes past the bounds of the variable buffer, allowing an attacker to change the values of other data stored in memory. A buffer overflow can occur in several places in memory. The most common—and easiest for an attacker to exploit—are buffer overflows that occur on the stack, the location in memory where local variables and return pointer information are stored. Overflows can also occur in other areas of memory such as the heap, but these are generally more difficult to exploit because minor changes in the application's execution environment can change the location of data in memory.

Many functions in the C and C++ programming languages (a sampling of the common offenders is shown in Table 8.1) do not do bounds checking on data. These functions begin to place data in memory at the starting address of the target buffer but continue to store the data in subsequent addresses, essentially ignoring the bounds of the buffer. If the space set aside to store this data is on the stack, the result is that data can overwrite the return address of the current function or subroutine. When this happens, an attacker can take control of the application. In the sections that follow, we take a closer look at stack and heap overflows.

TABLE 8.1 Sampling of C and C++ Functions Prone to Buffer Overflows When Given Untrusted Data

Some Common Functions in C and C++ Vulnerable to Buffer Overflows	
gets()	vsprintf()
sprintf()	fscanf()
strcat()	Scanf()
strcpy()	getopt()
streadd()	getpass()
strtrns()	fread()
index()	realpath()

STACK OVERFLOWS

Stack overflows are the most exploited class of buffer overflows because they are the easiest to exploit. The concept of a stack buffer overflow is that data placed on the stack overflows the space allocated for it, often overwriting the return address of the current function in memory. The *return address* is the address at which the application begins executing instructions once the current function is done executing. After the function returns, the value in the return address is placed in the EIP (Extended Instruction Pointer) register, which holds the memory address of the next instruction to be executed.

ON THE CD

To understand how stack overflows work let's take a look at a specific example. The source code that follows is taken from HexDump, an application included on the CD-ROM for illustrative purposes at Source Examples\Chapter 8\Listing 1-hexdump.cpp.

```
int main(int argc, char* argv[])
{
  .
  .
  .
  FILE* fp = fopen(fileName, "rb");

  if (!fp)
  {
    MessageBox(NULL, "Invalid file", "Error", MB_ICONSTOP|MB_OK);
    return 0;
  }
```

```
      DoHexDump(fp);
  .
  .
  .
}

void DoHexDump(FILE* file)
{
  if (file != NULL){
  int len;
  unsigned char data[512]; //Buffer declared to hold the data in the
file
  FILE* fp = file;

  fseek(fp, 0, SEEK_END);
  len = ftell(fp);
  fseek(fp, 0, SEEK_SET);
  fread(data, 1, len, fp); //reads the file data and places
//it into the data variable
  fclose(fp);
  DumpHex(data, len); //function to output the data
  }
  else{
      printf("Failed to open file!");
  }
}
```

When run, main calls the DoHexDump function, which in turn calls the function DumpHex. After the DumpHex function is executed, the application then attempts to exit from the DoHexDump function. When the DoHexDump function is called from main, the return address is pushed onto the stack. This address points to the instructions in main to be executed after DoHexDump has finished executing.

Within the DoHexDump function, the array data is declared as 512 bytes. After the array is declared, the data stack of the application likely looks like Figure 8.2.

Depending on how the stack is ordered, if we read in more than 512 bytes from the file, the 513th through 516th bytes [1] are going to overwrite the return address of the function DoHexDump. The result is that the application executes normally (provided other critical data was not overwritten also) until it reaches the end of the DoHexDump function. Once this function is done executing, the application attempts to return control to main by placing the return address (which was saved on the stack when the DoHexDump function was called) into the EIP register. In the 32-bit x86 architecture, the Extended Instruction Pointer (EIP) register holds the memory

Buffer Overflow Vulnerabilities 111

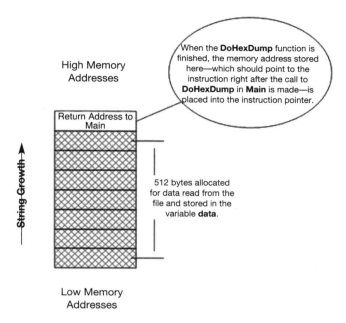

FIGURE 8.2 Data stack after array is declared.

address of the next instruction to be executed. When a function returns, the return address is popped into the EIP register, which tells the application the location of the next instruction is to execute.

Going back to the HexDump example, if we were to create a file containing 516 "a" characters (512 to fill up the allocated buffer for data and then 4 to overwrite the return address), we would overwrite the return address of the function DoHex-Dump with the value 61616161 (note that 61 is the ASCII value of "a"). This means that this value would be placed into the EIP register after the DoHexDump function is finished executing, and the application would then try to execute the command stored at memory address 61616161, which contains no executable instructions (see Figure 8.3).

In Windows XP SP1 the result is an application crash (see Figure 8.4).

If we run HexDump under a debugger (Figure 8.3), we can see that the reason it crashed is because it is trying to execute instructions at address 61616161 where there are no instructions. What we have just described is a stack-based buffer overflow. If an attacker can gain control over the instruction pointer (EIP), then he can manipulate it to execute instructions of his choice, which are usually also contained

112 The Software Vulnerability Guide

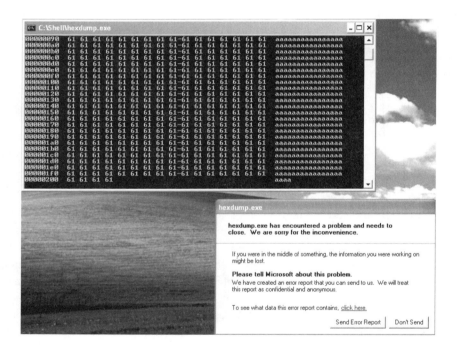

FIGURE 8.3 The EIP has been overwritten with "61"s, the ASCII equivalent of "a"s.

FIGURE 8.4 The application then tries to execute the instructions at address `61616161` and promptly dies.

in the data that causes the overflow to occur. Buffer overflows can also occur in the heap. Many heap-based overflows (that will be discussed later) are also exploitable using different techniques. The implications of having an exploitable buffer overflow in an application that accepts data from untrusted users are dire. An attacker can often gain complete control over a machine.

If a stack overflow occurs with random data, the application is likely to crash because some data that the application depends on is altered randomly by a user. Therefore, a crash after entering a long data string into a data field, API parameter, file field, or network parameter is a good indicator of a buffer overflow vulnerability. In some cases, however, buffer overflows can have more subtle symptoms, such as application instability or the corruption of other variables.

EXPLOITING STACK OVERFLOWS

Many stack overflows are exploitable by inserting enough data into the buffer so that the return address of the current function is overwritten. The strategy is to change the value of the return address so that it points to other user-supplied data on the stack. An attacker can then enter data values into the application that are interpreted as machine instructions. These instructions are commonly referred to as either *shellcode* or *op-code*. In our running HexDump example, we now illustrate how an attacker might take advantage of this buffer overflow. Our goal is to make HexDump launch Microsoft Notepad on the local machine. This is not something the application was designed to do. We are faced with the challenge of manipulating the application's execution through data.

We already know that we can overwrite the return address of the function Do-HexDump and thus ultimately control the value of EIP. Our next step is to find a location in memory that contains some more of our data; data that we can then point EIP to that essentially tells the application to interpret this data as machine code. After an overflow has occurred, the most common place to look for our data in memory is at the top of the stack. The Extended Stack Pointer (ESP) register always points to the top of the stack. If we look at the application in the ntsd debugger, we can see that data contained in the text file—in this case a bunch of "a"s—is indeed at the top of the stack, as shown in Figure 8.5.

By changing the values in the text file to a non-repeating character like "abcdefghi..." it is fairly easy to determine which "a"s in our file are the ones at the top of the stack. In this case, they are the ones at position 517, 518, and so on, right after the "a"s that overwrite the return address.

Our next step is to find a function that we can call that executes a command string so we can launch Notepad. HexDump loads the MSVCRT.DLL library, which exports a function called system that can do the trick. In cases where we need functions from other operating system libraries that are not loaded by the application,

```
ntsd hexdump bigfile.txt

Access violation - code c0000005 (first chance)
eax=0000004c ebx=7ffdf000 ecx=7c02d0e7 edx=7c04b0e0 esi=0006f38c edi=00000000
eip=61616161 esp=0011fb5c ebp=0012ffc0 iopl=0         nv up ei pl nz na pe nc
cs=001b  ss=0023  ds=0023  es=0023  fs=0038  gs=0000              efl=00010202
61616161 ??                     ???
0:000> db esp
0011fb5c  61 61 61 61 61 61 61 61-61 61 61 61 61 61 61 61  aaaaaaaaaaaaaaaa
0011fb6c  61 61 61 61 61 61 61 61-61 61 61 61 61 61 61 61  aaaaaaaaaaaaaaaa
0011fb7c  61 61 61 61 61 61 61 61-61 61 61 61 61 61 61 61  aaaaaaaaaaaaaaaa
0011fb8c  61 61 61 61 61 61 61 61-61 61 61 61 61 61 61 61  aaaaaaaaaaaaaaaa
0011fb9c  61 61 61 61 61 61 61 61-61 61 61 61 61 61 61 61  aaaaaaaaaaaaaaaa
0011fbac  61 61 61 61 61 61 61 61-61 61 61 61 61 61 61 61  aaaaaaaaaaaaaaaa
0011fbbc  61 61 61 61 61 61 61 61-61 61 61 61 61 61 61 61  aaaaaaaaaaaaaaaa
0011fbcc  61 61 61 61 61 61 61 61-61 61 61 61 61 61 61 61  aaaaaaaaaaaaaaaa
0:000>
```

FIGURE 8.5 A look at the memory pointed to by ESP (the Extended Stack Pointer) reveals that data from the input file—in this case a bunch of "a"s—is at the top of the stack.

we would need to load the libraries ourselves using the KERNEL32.DLL function LoadLibrary.

We now need a space in memory to store our command string notepad.exe. The memory address of that string is then passed to the system function. To find the address of the system function we first look up the address of that function in the MSVCR70.DLL library. In the ntsd debugger this is done with the command:

```
x msvcr70!system
```

which gives the address 0x7c021de4 (Figure 8.6). It would be unwise for us to call this address directly because if this library is provided by the OS or a third party, the library might be updated with a patch or service pack and the address of system is likely to change. Instead, we call the function through a pointer to this address, which is stored in our module. ntsd shows that our module is loaded in the range 0x00400000–0x00404000. The command for this in ntsd is:

```
s 00400000 00404000 e4 1d 02 7c
```

which returns the address 0x00402030.

It is fairly trivial to find a spot in memory to store our string given the sea of "a"s on top of the stack. We've decided to put them at offset 545 in our text file, which ends up 545 − 517 = 28 decimal (1C hex) away from the top of the stack. Adding 1C to the address of ESP at the time of the overflow gives the address 0x0011fb5c + 0x0000001c = 0x0011fb78.

Buffer Overflow Vulnerabilities 115

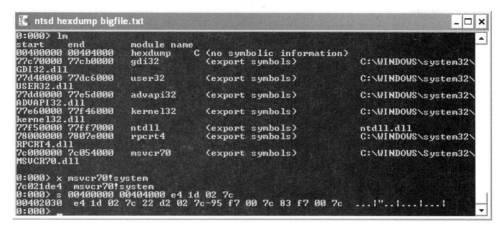

FIGURE 8.6 The ntsd debugger reveals the address of the `MSCVR70.system` function.

Now that we have the address of the string `notepad.exe` and the address of the `system` function, we want to execute the following code:

SHELL CODE	ASSEMBLY	DESCRIPTION
68 78 fb 11 00	Push 0x0011fb78	Push the location of the string `notepad.exe`.
FF 15 30 20 40 00	Call Msvcr70.System	Call the system function in `msvcr70.dll`. In this case we are calling indirectly through a pointer.

Our attack file now contains the address of ESP at offset 513–516 followed immediately by our shellcode. The final file is shown in Figure 8.7.

When we launch HexDump.exe and pass in the filename of our file through the command line, HexDump launches Notepad.

HexDump as well as our exploit is included on the CD-ROM in the Source Examples\Chapter 8 folder. We encourage you to play with the values in the text file to change the behavior of this application.

What we've just walked you through is the process that an attacker might follow to exploit a stack overflow. Understanding how an attacker views our software places us in a much better position to defend our applications and find these types of problems through testing. In the next section we will take a look at buffer overflows that occur in the heap.

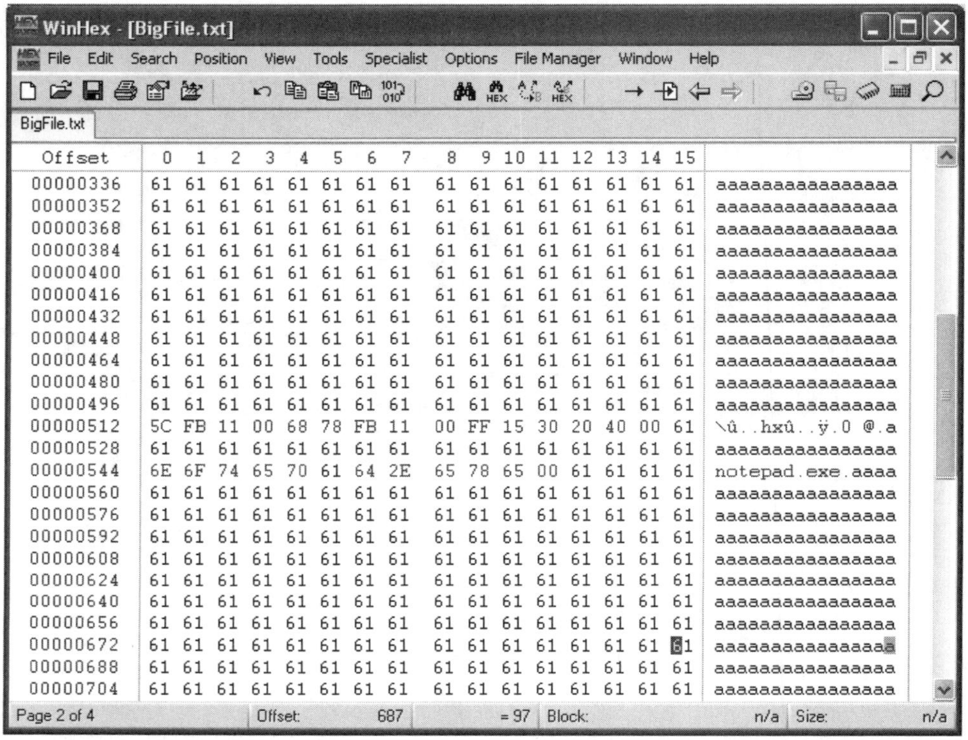

FIGURE 8.7 WinHex shows our final exploit file. The value that overwrites the return address is at offset 512, followed immediately by our shellcode.

HEAP OVERFLOWS

Buffer overflows can also occur in the application *heap*. The heap is an area of memory where storage is dynamically allocated and freed during execution as necessary. The heap is a very awkward memory structure due to its dynamic nature. It is filled with interspersed blocks of allocated and free memory, as shown in Figure 8.8. Various platforms and compilers keep track of heap data in different ways. Typically, each block of data is encased with headers that point to the next allocated block of memory in the heap using a linked list. In some architectures, the free space interspersed between allocated blocks also contains a linked list structure pointing to the next free block of memory. When memory is allocated or freed, the pointers contained in these structures are updated appropriately.

Buffer Overflow Vulnerabilities   117

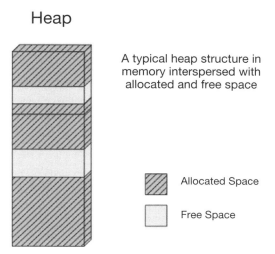

FIGURE 8.8 An illustration of what a typical heap structure looks like.

In many instances using heap variables is essential, namely when the size of data to be read from a user cannot be specified statically. A good example would be reading data from a file. Here, we might need to declare the size of the variable to hold the file data based on some runtime properties (such as the size of the file). Heap variables can be declared by using the new operator:

```
char *pCharArray = new char[256];
```

The preceding statement allocates 256 bytes of memory in the heap, at the address stored in pCharArray. The same result can be accomplished using the C function malloc, as shown here:

```
char *pCharArray = (char *)malloc(256);
```

Applications are made vulnerable to heap overflows by using the same sorts of unsafe string manipulation functions that cause stack overflows (see Table 8.1). When an overflow in a heap variable occurs, typically some of the linked list structure that keeps track of allocated and free memory blocks is corrupted, and the results can be unpredictable. Sometimes, the result is that data in a contiguous block is overwritten and application execution continues normally. In other instances the application might crash as it references the corrupt structure.

The code listing that follows demonstrates a heap overflow. The variables `buffer1` and `buffer2` (both on the heap and both allocated to be 16 bytes) are initially filled with "A"s. The user is then prompted to enter text and that text is put into `buffer1` using the `gets` function. (As we warned the reader in Chapter 1 about `strcpy`, we should point out that `gets` is unsafe and should never be used.) `gets` does not do bounds checking on the destination buffer, and thus if a user enters more than 16 characters, user data is written past the bounds of the buffer and corrupts data immediately after the buffer in the heap.

```c
#include <stdio.h>
#include <stdlib.h>
#include <string.h>
#include "stdafx.h"
#define BUFSIZE 16


int main(int argc, _char* argv[])
{

    char *buffer1 = (char *)malloc(BUFSIZE);
    if (buffer1 == NULL)
      return 0
    char *buffer2 = (char *)malloc(BUFSIZE);
    if (buffer2 == NULL)
      return 0
    memset(buffer1, 'A', BUFSIZE-1);
    buffer1[BUFSIZE-1] = '\0';

    memset(buffer2, 'A', BUFSIZE-1);
    buffer2[BUFSIZE-1] = '\0';

    printf("buffer1 pointer = %p, buffer2 pointer = %p", buffer1, buffer2);
    printf("\n\nValue in buffer1: %s", buffer1);
    printf("\nValue in buffer2: %s", buffer2);

    printf("\n\nEnter new value to be placed in buffer1 with gets(): ");

    gets(buffer1);

    printf("buffer1 pointer = %p, buffer2 pointer = %p", buffer1, buffer2);
    printf("\n\nValue in buffer1: %s", buffer1);
    printf("\nValue in buffer2: %s", buffer2);
```

```
        printf("\n\nPress enter to (try) and free buffer1...");
        getchar();
        free(buffer1);

        printf("\n\nPress enter to (try) and free buffer2...");
        getchar();
        free(buffer2);

        return 0;
}
```

We compiled this code with Microsoft Visual C++ .NET. Figure 8.9 shows that buffer1 is located at address 0x00320850 and buffer2 is at address 0x00320890. Figure 8.9 also illustrates that when we enter 15 "c"s (the last character in the allocated memory is automatically set to NULL to indicate the end of a string) as input, the application stores them appropriately in buffer1, overwriting the "A"s. The two buffers are then freed, and the application exits normally because no overflow has occurred.

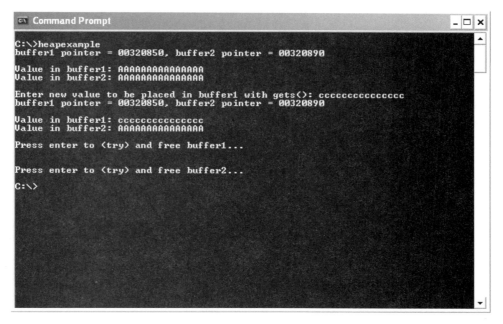

FIGURE 8.9 Our compiled heap example is run and reveals the addresses of the two buffers on the heap. When we input 15 characters into the 16-character buffer, no overflow occurs.

Next, we rerun the application, this time entering 20 "c"s. The gets function blindly places our data into buffer1 and writes past the 16 bytes originally allocated. Figure 8.8 shows that the application indeed returns the 20 "c"s to us when we examine the contents of the buffer. That's because when a string is referenced, many functions—in this case printf—continue to read the string until they read a NULL byte (0x00). Because the NULL byte was placed after the 20 characters, all characters are read and displayed. Also, note that the data held in buffer2 is still the original "A"s.

Where was the extra data put? When gets overwrote the bounds of the buffer allocated to store that data, information on the heap that keeps track of block allocations was corrupted. When our application then tries to "free" that buffer, an exception is thrown, as shown in Figure 8.10. Because it is a first chance exception, we can choose to ignore it, and when we do, the application exits normally.

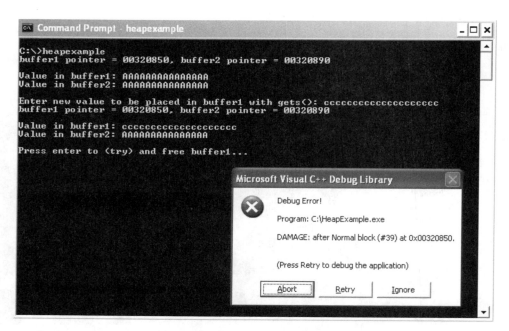

FIGURE 8.10 Our overflow corrupts the heap structure that causes an ignorable first chance exception.

Let's take a look at the addresses that these buffers are stored at in the heap. From Figure 8.9, buffer1 starts at address 0x00320850 and buffer2 starts at address 0x00320890. The variable buffer1 is allocated to be 16 bytes, but the difference between the two start addresses is 40 hex, which translates (in decimal) into 64 bytes. If you use a different compiler and/or run the application on a different platform, your results might be different. Why the extra space? Some of this space is used for housekeeping information about the heap. In applications that dynamically allocate and free memory multiple times during execution you might notice seemingly random gaps between the address of heap variables. Aside from keeping linked list type information on what areas of the heap are free or allocated, areas of the heap are allocated based on the availability and size of contiguous free space. In this particular instance, we have 64 bytes until our data starts to overwrite buffer2. If we enter 65 "c"s into this application, the result is that buffer2 is overwritten with 1 "c", as shown in Figure 8.11.

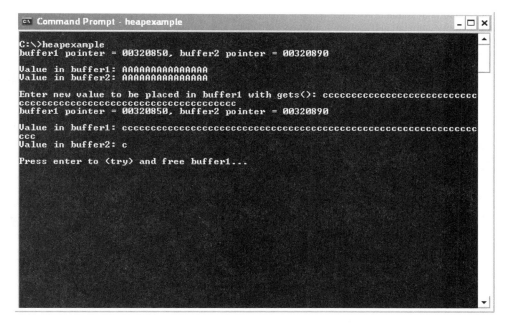

FIGURE 8.11 When we feed the application 65 "c"s, buffer2's data is overwritten with a single "c".

Again, because of the way gets (a function that is inherently insecure and should never be used) works, a NULL byte is placed immediately after the last "c" in memory and thus our printf function reads the variable buffer2 as containing only one letter. If we take a look at the memory at the starting address of buffer2 (Figure 8.12), we can see that the original "A"s are still there in memory after the NULL (0x00) byte.

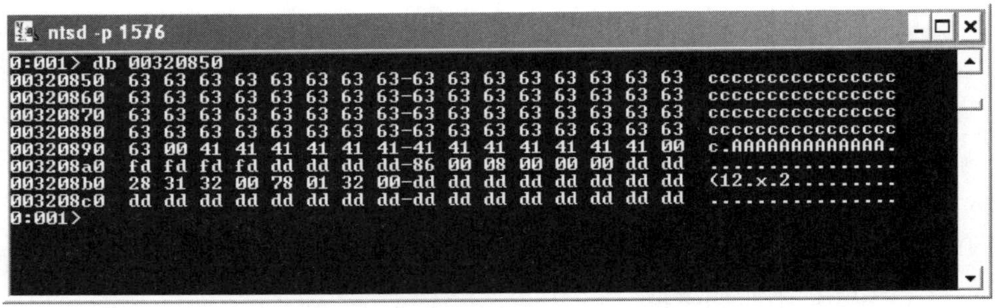

```
 0:001> db 00320850
 00320850  63 63 63 63 63 63 63-63 63 63 63 63 63 63 63 63  cccccccccccccccc
 00320860  63 63 63 63 63 63 63-63 63 63 63 63 63 63 63 63  cccccccccccccccc
 00320870  63 63 63 63 63 63 63-63 63 63 63 63 63 63 63 63  cccccccccccccccc
 00320880  63 63 63 63 63 63 63-63 63 63 63 63 63 63 63 63  cccccccccccccccc
 00320890  63 00 41 41 41 41 41-41 41 41 41 41 41 41 41 00  c.AAAAAAAAAAAAA.
 003208a0  fd fd fd fd dd dd dd-86 00 08 00 00 00 dd dd     ................
 003208b0  28 31 32 00 78 01 32 00-dd dd dd dd dd dd dd dd  (12.x.2.........
 003208c0  dd dd dd dd dd dd dd-dd dd dd dd dd dd dd dd dd  ................
 0:001>
```

FIGURE 8.12 If we take a look at what's actually stored in memory, we see that only the first 2 bytes of `buffer2` were overwritten.

Freeing `buffer1` causes several first chance exceptions that we can ignore. Attempting to free `buffer2`, however, results in a crash.

Several methods exist to exploit the way that `free` is implemented in many compilers to use a heap overflow to eventually take control of an application by forcing it to execute shellcode data supplied by the user. The next section talks about some of the methods used to exploit non-stack buffer overflows.

EXPLOITING BUFFER OVERFLOWS: BEYOND THE STACK

Earlier in this chapter we illustrated a basic stack exploit. However, many other types of buffer overflows can be exploited using different techniques. Consider the following source code listing:

```
#include <stdio.h>
#include <stdlib.h>
#include <string.h>
#include "stdafx.h"

int main(int argc, char **argv)
{
    FILE *LogFileHandle;
        static char buffer[16], *LogFile;

        LogFile = "logfile.txt";

    printf("\nPointer to ARGV[1] = %p", argv[1]);
    printf("\nPointer to ARGV[2] = %p", argv[2]);
    printf("\nPointer to buffer = %p", buffer);
```

```c
        printf("\n\nPointer to LogFile is stored at %p", &LogFile);
        printf("\nValue of the LogFile pointer is %p", LogFile);

        printf("\n\n\nWARNING! Only the file %s can be updated.\n",
LogFile);
            printf("%s, please enter data to be placed into file %s: ",
argv[1],    LogFile);
            gets(buffer);

        printf("\nPointer to ARGV[1] = %p", argv[1]);
        printf("\nPointer to ARGV[2] = %p", argv[2]);
        printf("\nPointer to buffer = %p", buffer);

        printf("\n\nPointer to LogFile is tored at %p", &LogFile);
        printf("\nValue of the LogFile pointer is %p", LogFile);


            printf("\n\n\nThank you. Log file %s will be updated.\n",
LogFile);


            LogFileHandle = fopen(LogFile, "w");
            if (LogFileHandle == NULL)
            {
                fprintf(stderr, "Log file %s currently unavailable.\n",
LogFile);
                exit(-1);
            }

            fputs(buffer, LogFileHandle);
            fclose(LogFileHandle);
}
```

This application illustrates a commonly exploitable buffer overflow with applications written in C. Here we have an application that, when run, asks the user to enter some data that is placed into a log file named logfile.txt. The string logfile.txt is stored in memory as a static variable with the pointer LogFile holding the address to that string. The intent is for the user to not be able to alter the filename.

The variable buffer is declared to be 16 bytes and is created to store the input provided by the user. This is declared as a global variable, and thus it is not stored on the heap. Instead, we have the 16-byte buffer in memory, followed immediately by the value of the LogFile pointer.

124 The Software Vulnerability Guide

> *We compiled the source in Visual Studio.NET 2003 and the sample output shown was produced by the resulting executable.*

We can see from Figure 8.13 that when we run the application, storage for the variable `buffer` begins at address 0x0042b500 and the pointer `LogFile` is stored at memory address 0x0042b58, which is 0x8 = 16 bytes after the beginning of buffer.

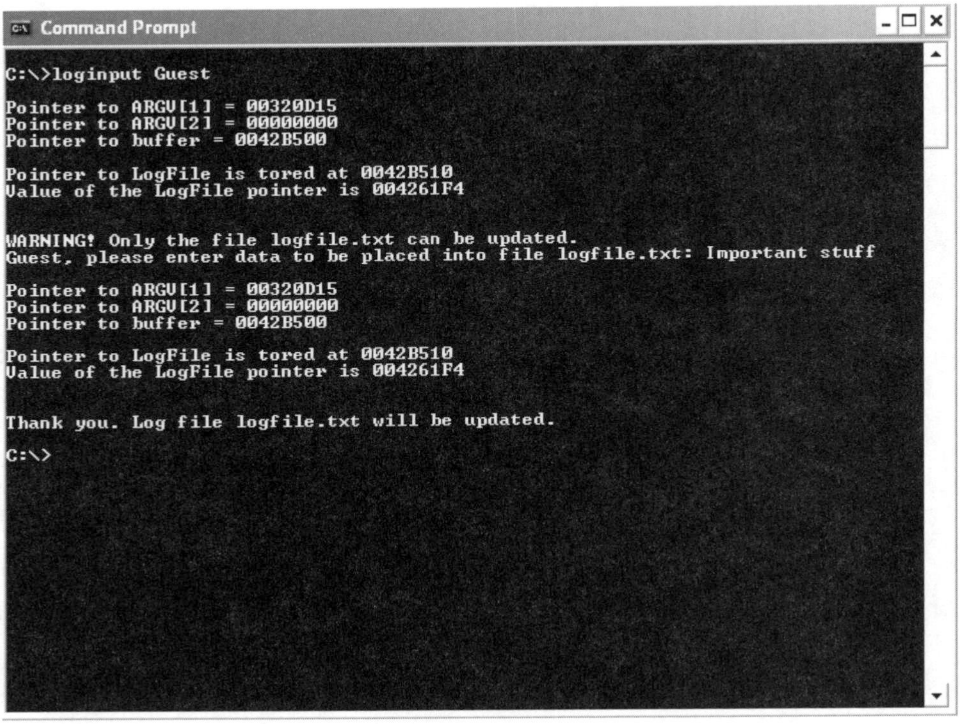

```
C:\>loginput Guest
Pointer to ARGV[1] = 00320D15
Pointer to ARGV[2] = 00000000
Pointer to buffer = 0042B500
Pointer to LogFile is tored at 0042B510
Value of the LogFile pointer is 004261F4

WARNING! Only the file logfile.txt can be updated.
Guest, please enter data to be placed into file logfile.txt: Important stuff
Pointer to ARGV[1] = 00320D15
Pointer to ARGV[2] = 00000000
Pointer to buffer = 0042B500
Pointer to LogFile is tored at 0042B510
Value of the LogFile pointer is 004261F4

Thank you. Log file logfile.txt will be updated.

C:\>
```

FIGURE 8.13 Running our compiled application reveals the memory addresses where our buffers are.

When we run the application, it prompts the user for the data to be written to `logfile.txt`. The problem, however, is that the space allocated to store this data is only 16 bytes long. If we enter 20 "a"s into this field, we end up overwriting the value of the pointer `LogFile` with `61616161` (where `61` is the ASCII value of the letter "a"). This also has the side effect of crashing the application when the program tries to read data from this bogus memory address (0x61616161), as shown in Figure 8.14.

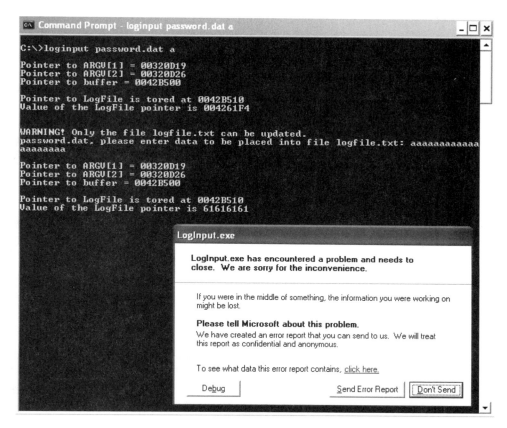

FIGURE 8.14 If we enter more than 16 characters, the LogFile pointer is overwritten with the hex values of the 17th–20th characters (in this case "a"s).

To exploit this ability to overwrite the LogFile pointer we are going to use the fact that the application reads command-line parameters. The first parameter that we feed to the application is intended for the user's name, and it is stored in argv[1]. The second command-line parameter (which isn't used by the program) is stored at argv[2], the third at argv[3], and so on. Our strategy here is to change the file that the application writes to. We are going to attempt to force it to use the filename password.dat.

An interesting strategy is to enter the name password.dat as a command-line argument. This value is stored on the heap. When we execute the application with this parameter, it is stored at address 0x00320d15. Ideally we would now like to use the buffer overflow discovered earlier to change the value of the pointer LogFile from 0x004261f4 (the address of `logfile.txt`) to 0x00320d15 (the address of `password.dat`). To do this, when prompted, we need to enter a string of 16 characters followed by the ASCII characters representing 0x15, 0x0d, and 0x32. Note that the byte values are entered in reverse order because addresses in the x86 architecture are stored with the least significant byte first, a convention called little endian. Because the `gets` function places a 0x00 at the end of a string, this means that we would overwrite the `LogFile` pointer so that it now points to `argv[1]` where we have the string `password.dat`. The bytes that make up the address—0x15, 0x0d, and 0x32—correspond to the characters CTRL-U, CTRL-M, and "2", respectively. This presents a challenge to the attacker because CTRL-M, the newline character, would terminate our string. To get around this, we can increase the size of `argv[1]` by inserting characters and thus move the value of `argv[2]` to a memory location that can be entered more easily through command input. In this case, 231 "a"s in `argv[1]` makes the address of `argv[2]` on the heap 0x00320e01. This means that we must enter the ASCII values for 0x01, 0x0e, and 0x32, which are CTRL-A, CTRL-N, and "2". The result of entering these values in positions 17, 18, and 19 of the input string is shown in Figure 8.15.

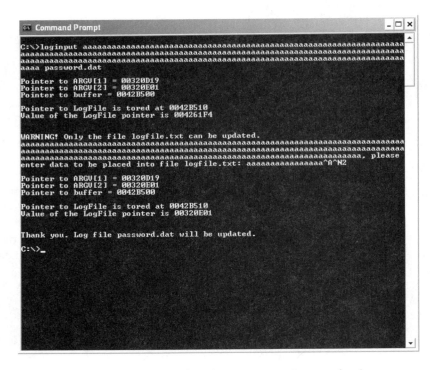

FIGURE 8.15 We can manipulate the `LogFile` pointer so that it now points to the string `password.dat`.

We have now changed the value of the `LogFile` pointer so that it references the second parameter we entered through the command line. If the process has the appropriate permissions, the result is that an attacker can now force the application to overwrite any file on the system.

Attackers have used techniques like this to escalate their privilege over a system. Other attacks that exploit overflows elsewhere in memory are also possible, and if a way to exploit a buffer overflow in a popular commercial application exists, some attacker, somewhere, will eventually find it.

> **FAMOUS FAILURES: THE $2 BILLON BUFFER OVERFLOW**
>
> In July of 2001, a new worm dubbed Code Red was identified as spreading rapidly on the Internet. The worm affected Microsoft Windows machines running the Internet Information Server 5 (IIS5). IIS (as the default Web server shipped with Windows server operating systems) is a widely used platform to host Internet sites. Prior to the outbreak of this worm, many ordinary desktop systems also had IIS turned on by default. The worm exploited a buffer overflow in the Indexing ISAPI of IIS and used it to take control of its victims. It used this control to deface the Web sites running on that machine and made a failed attempt at harnessing infected machines to attack the Web site of the White House. Within days, variants of the worm emerged and the infection spread rapidly to vulnerable hosts. The result was disastrous. Many corporate networks ground to a halt as traffic from the worm trying to propagate consumed bandwidth. The cost of lost productivity and cleanup is estimated at $2.6 billion [Ryder02].

FINDING THIS VULNERABILITY

As an industry, we've learned a lot about buffer overflows in the past five years. Scores of tools have been written to help testers and developers find these problems earlier in the application development lifecycle with some success. Still, buffer overflows continue to dominate public vulnerability databases in not just old code but new applications as well. You can use several approaches to find buffer overflows in software. Given that this is a longstanding problem and that a fair amount of people have tried to address it, we've broken this section up into code-based tools and techniques and techniques that testers can apply on the running executable to track these problems down.

White-Box Testing Techniques and Tools

From a white-box perspective, several functions in C and C++ can easily lead to buffer overflows and should be replaced by their safer alternatives. In Table 8.1 we presented some of the more common vulnerable APIs and functions in the C runtime. A hand audit of the code for these functions is certainly an option but can be time consuming. Once you've located one of these functions, it can be very difficult to then track backwards through the code to find out if this code-level weakness actually translates into an application vulnerability. If, for example, data is truncated before it is passed into a function that used a `strcpy`, then you have no immediate vulnerability. We've heard the truncation argument numerous times for developers, but the larger issue is that code changes and later another module, written by another developer, might call into the risky function. In general, it is cheaper to fix a potential buffer overflow once it's found than to fix it post release when someone finds a route to exploitation.

Several automated source-scanning tools for C can make the process of searching through your source code for vulnerable functions easier. RATS, the Rough Auditing Tool for Security, is a free source code scanner produced by Secure Software (*www.securesw.com*) under active development that is capable of scanning C and C++ source code for functions vulnerable to buffer overflows. The ITS4 security scanner by Cigital, Inc., (*www.cigital.com*) is also free and can be used to scan C and C++ for related problems. Flawfinder (*www.dwheeler.com/flawfinder*) is another GPL vulnerability finder that scans C and C++ code for a variety of security problems including buffer overflows.

The biggest problem with using automated scanners is the amount of "noise" or false positives they generate. Practitioners should be sensitive to the fact that not all issues identified by scanners are exploitable, but again the rule of thumb should be to nip potential buffer overflow issues in the bud.

Black-Box Testing Techniques and Tools

The first step to finding these issues in the running application is to identify an interesting input string as a target. From a security perspective, look for data that is gathered from an untrusted user through Web interfaces, data files, API interfaces, or the registry. Particular care should be taken when dealing with applications that run at a higher privilege than their users. In these cases, input from the GUI must be scrutinized. Successful attacks here would allow a user to execute commands at the enhanced privilege level of the running application.

Once you have identified a target, the technique is fairly straightforward: apply long strings. When looking for this type of vulnerability, though, you must be cognizant of any input filters that might be bypassable by an attacker. A common example is the client-side filtering that many Web applications do to constrain data. An attacker, however, can easily bypass this control by posting data directly to the server. The same is sometimes true of standalone applications. Several techniques can bypass user interface filters, and testers should always be vigilant of alternate ways to deliver data.

Input files, registry data, and API parameters and return values can also be a source of potentially harmful input. For a file that contains a series of numerical data that the software reads, for example, you might want to use a text editor and include letters and special characters. If successful, this attack usually results in denial of service by either crashing the application or bringing down the entire system. We have found random file corruption to be very effective at exposing buffer overflow problems in file parsing code.

At a high level, you need to be careful to check three interfaces for long string inputs:

Network interface: If your application processes data that comes from a remote source via some communications protocol, that input stream needs to be carefully checked for long strings. The Nimda, Code Red, and Slammer worms were all enabled over a network interface by vulnerable code that failed to constrain strings embedded in the protocols.

Filesystem interface: Files can act as proxies for a remote user. Many viruses have spread through executable files and script files but surprisingly few have taken advantage of a buffer overflow in the application used to create them. Our own testing has found that an amazing number of widely used applications are susceptible to buffer overflows in the files they read. BugTraq is loaded with examples, as are many other public vulnerability databases. The reason why buffer overflows are so concerning here is that it would be very difficult to identify a file as potentially malicious until it's too late. Random file corruption is particularly effective here at exposing buffer overflow issues. The idea behind random corruption is fairly simple: add long strings at random locations within files. Admittedly, on several file types this techniques is ineffective (fields that have specified lengths and are not delimited, for example), but our own testing shows that this technique is effective a surprising number of times.

Programmatic interfaces: API calls to the application can contain long strings in parameters, and thus these arguments must be validated by the application under test. Also, API calls made by your application can contain long strings through "in" parameters, return values and data pointed to by parameters. All such data needs to be validated before it is used, and from a testing perspective, this can provide a rich testing surface.

FIXING THIS VULNERABILITY

C and C++ are powerful languages, and as a result, they offer the developer a tremendous amount of power to manipulate system resources and memory. That power, however, also makes it fairly easy to make mistakes and poor assumptions when handling data. Special care must be taken when copying, moving, or storing data in these languages. Any assumption regarding size must also be enforced in code. Many buffer overflow issues can be resolved by replacing known dangerous C and C++ string manipulation functions with their safer alternatives. Table 8.2 lists the common offenders and their safer, more constrained replacements.

TABLE 8.2 Unsafe C Routines and Safer Alternatives

Function	Description	Safer Alternative	Description
gets	Takes input from stdin stream until it gets a linefeed or carriage return. It blindly stores this data in the specified buffer.	fgets	Using the fgets function, you can specify the length of data received and stored.
strcpy	Copies data blindly from a source buffer to a destination buffer.	strncpy	Allows a user to specify the amount of data copied.
strcat	Concatenates two strings by copying and adding data from one buffer to another.	strncat	Concatenates two strings with length specified.
sprintf	This function takes the contents of one buffer, formats it and stores it in a destination buffer.	snprintf	Formats the data from one buffer and places it in another but allows the developer to specify the size of data.
scanf	Reads formatted input from stdin. Can be dangerous if used with an unbounded specifier like %s.	scanf with bounded format specifier	If scanf is used with a format specifier that constrains length, then it can be safe.

Summary Sheet—Buffer Overflows

Problem:

Buffer overflows are by far the most reported and exploited security vulnerabilities in software. A buffer overflow occurs when the data passed to an application is larger than the space allocated to store that data in memory. Languages like C and C++ make it easy for developers to make bad assumptions about the length of data supplied to an application. Sometimes, if the data supplied is larger than the memory allocated to store it, that data can overwrite other areas in memory, possibly allowing an attacker to execute arbitrary instructions. Buffer overflows can occur either on the stack or the heap. Stack buffer overflows are the most common, and many are easy to exploit. Heap overflows, however, are a serious concern and can also be exploitable.

Potential Impact:

Execution of arbitrary commands. Complete control of an application through user-supplied data. Catastrophic failure.

Habitat:

Buffer overflows can occur in applications written in languages that do not enforce bounds and type checking. Because C and C++ fall into this category and most software is written in those languages, the problem is widespread. Even applications written in "safe" languages such as Java and C# might be at risk because of external libraries called from those applications, which might be written in C or C++.

Tools You Need to Find It:

Typically, the application UI is a good place to start to enter long strings (no tools needed). For buffer overflows that are exploitable through the filesystem, network, or APIs, several "fuzzing" tools can help. See Chapter 9 for more information.

How to Look for It:

Feed long strings into data fields. A data field, however, might be through the user interface, the filesystem, an API, or the network. Arbitrary long strings might not be effective, however. The data might have to be in a certain format or contain delimiters itself so that it is parsed into variables in a way to create the overflow condition. Because buffer overflows are arguably the most prevalent (and one of the most dangerous) classes of software vulnerabilities, Chapter 9 is dedicated to techniques that can be used to find these problems.

Symptoms of Failure:

The most common symptom of failure is an application crash after a long string has been entered into a field. The crash often results from user data overwriting the return address of a function in memory with garbage. Application instability and data corruption is also another common symptom as data from one field overflows into other data storage areas and corrupts that data.

Famous Failures/Exploits:

Two very famous buffer overflows include the 1997 MIME-conversion buffer overflow in sendmail and the 2001 buffer overflow in IIS indexing service, which led to the Code Red worm.

ENDNOTES

[1] This is dependant on several things, most importantly the compiler used. In many compiled versions, the bytes that hold the return address might be different because of other data stored on the stack like the Base Pointer (EBP). The version shown in this example compiled with the Microsoft C++ compiler without the /GS flag, which would have made exploitation significantly more difficult.

REFERENCES

[Ryder02] Ryder, Josh. "Castles Built on Sand: Why Software is Insecure." *Security Focus*. Online at *www.securityfocus.com/infocus/1541*. January 30, 2002.

9 Proprietary Formats and Protocols

In This Chapter

- Description
- Using "Fuzzing" to Find Vulnerabilities in File Formats and Protocols
- Preventing Problems with Proprietary Formats and Protocols

We don't give very much thought to the format of data stored or transmitted by an application. This is because, as users, we rarely interact with data in any sort of raw form; files and network communications are manipulated by applications—we open a Word file using Word, an HTML file using a browser, etc. Relying on applications to store and manipulate data in their own way is a fundamental concept of computing, one that has been preserved on as many varied media formats as punched cards and paper tape, magnetic disks and optical storage. Some changes over time, including the introduction of ASCII and then Unicode to define the binary representation of ordinary text, as well as XML, which is attempting to standardize the meta-format of data files, have meant that programmers have had less freedom in the past to invent their own representations of information. However, it is still pretty much up to the programmer to define his own format for storing and retrieving data. He might select formats that ease interoperability both with other applications and later versions of his own application, maximize readability, provide for protection from would-be attackers and errors, or minimize space and retrieval time. Too often, however, selection of storage and transmission formats is made based on other criteria: ease of programming, at the expense of readability or interoperability; close resemblance to in-memory data structures, at the expense of protection from errors and malicious modifications; or proprietary obscurity, to prevent interoperability with a competitor's product or to force upgrades to later versions.

Except for the occasional employment of encryption, security is not usually a factor in the design of a file format or protocol. This presents a problem, because though the application programmer might naively believe that only his program is able to generate, modify, or read the data associated with his file or protocol format, that is often not the case. An attacker can easily use techniques such as network fuzzing to generate or modify this data in a way that is harmful to the application and security. Additionally, an attacker with a basic knowledge of programming and an ability to interact with the application by means of a debugger can often reverse engineer the protocol or file format, meaning he can generate arbitrary files or even bypass some of the security measures inherent in them.

DESCRIPTION

File formats and protocols abound. A popular format description Web site for programmers contains links to 966 different file formats. Likewise, the IANA Port List, which documents assigned port numbers for protocols based on TCP and UDP only, contains over 13,000 lines. (For an up-to-date list of these port numbers, refer to *www.iana.org/assignments/port-numbers*). These give no real estimate of the number of file formats or protocols—in all likelihood the total number is close to the total number of programs ever written, a number in the millions.

With each of these formats are the associated routines required to encode and decode the format for use within the application. For well-defined, open protocols, this code can be reused; however, each application that uses a proprietary protocol likely has its own routines for encoding and decoding. These might be simple; if the protocol is plain text, as is the case with Telnet, the terminal protocol for Unix, and other timesharing systems, it simply passes the data on to other system functions intact. Likewise, a text or hex editor is generally able to manipulate any file by reading the raw bytes that make it up; no special processing of the file is required to make the data usable for these applications. Other formats can be mind-bogglingly complex. The (incomplete) documentation for the Word 8.0 format is half a megabyte; code that correctly manipulates all of the objects in this format would be decidedly larger in size. Complexity, as we have stated many times previously, is the enemy of security, so an implementer of a complicated format such as Word would need to employ a great deal more caution in implementing it.

What difference does format make when it comes to security? After all, a file is just a file, especially when it contains "ordinary" data with no expectation of confidentiality or privacy. Likewise, a network application that is supplied bad data likely just ignores that data, resets the connection, or processes it in a garbage-in, garbage-out fashion, right? In practice, this is not the case. The more a format or protocol is designed for obscurity, speed, or ease of programming, the less likely it is to contain the extensive error-handling needed to ensure that the application does not crash, expose a buffer overflow or format string vulnerability, or permit unauthorized operations. This is because programmers of proprietary protocols especially assume that only their application can read and write files in their format. If an attacker can force the application to accept a corrupt file, or a connection from a malicious application, all bets are off.

Same Data, Many Formats

Images are a great example of data that is represented in a variety of formats based on specific performance and compatibility factors. Figures 9.1, 9.2, and 9.3 show three representations of the same image file. The image (not shown) is a simple green arrow on a white background.

Figure 9.1 shows the image in the Windows bitmap (BMP) format. BMP is a basic binary format—it contains a small header, and each pixel is represented by a numeric value (1, 2, or 4 bytes depending on color depth) that describes its color.

The Encapsulated PostScript (EPS) format, shown in Figure 9.2, is a text format. EPS is designed to conform to the text-based PostScript language that is supported on a variety of printers and other devices. PostScript files are actually computer programs written in the PostScript language.

Figure 9.3 shows the same image in JPEG format, a portable, compressed image format. JPEG contains a mixture of text and binary data in the file.

Each of these formats represents different concerns. A program that parses the bitmap image format is not very likely to encounter many security concerns. Because the bytes of the file represent only colors, and every numeric value represents a valid color, any corruption of this file would likely result only in bad pixels in the rendered image. Apart from the dimensions of the image, there isn't really a "buffer" to speak of and handling overruns and underruns of data are relatively easy.

Encapsulated PostScript, on the other hand, presents an interesting set of problems. Because it is essentially a computer program, an attacker might sneak malicious commands into the file that the interpreter might execute. Infinite loops,

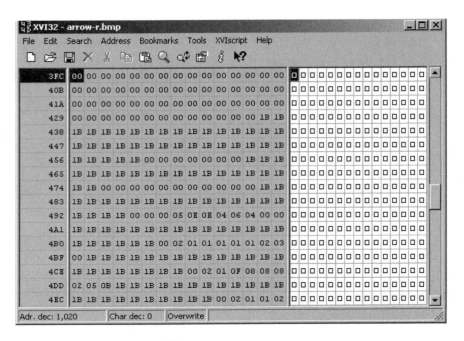

FIGURE 9.1 Arrow in BMP format.

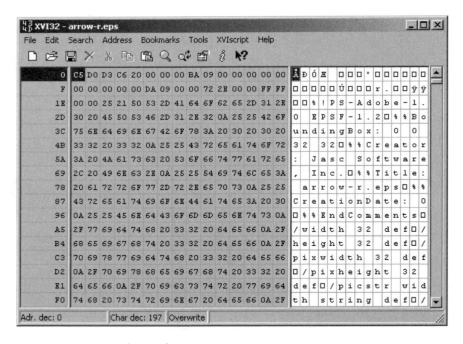

FIGURE 9.2 Arrow in EPS format.

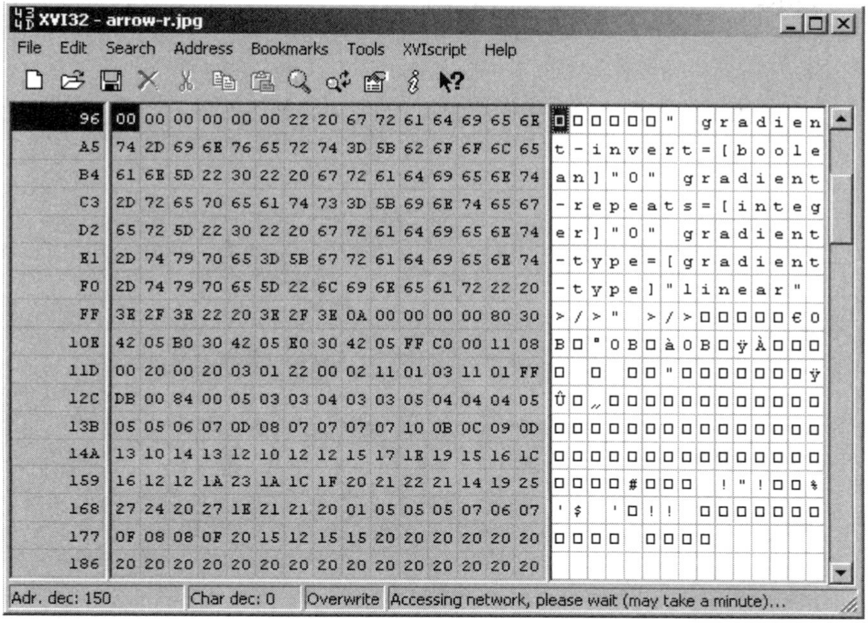

FIGURE 9.3 Arrow in JPEG format.

race conditions, and out of memory conditions caused by the code might result in a denial-of-service attack against the device or application that attempts to render the image. Finally, the language interpreter or file parsing routines might contain buffer overflows that are exploitable when very long command names or data element values are supplied.

Mixed text and binary formats, such as JPEG, present even more issues. Because the interpretation of a data element as text or binary is position-dependent, corruptions could lead to text being interpreted as image data, or vice versa. This kind of corruption often results in a crash that causes a denial-of-service attack and possibly an exploitable buffer overflow.

The Microsoft Clipart Gallery file format (CIL) contained just such a vulnerability, described in MS00-15. The CIL file format is used to download clipart from the Internet for use in Microsoft applications. In older versions of the application, a string that contained the destination filename of the compressed image was supposed to be bounded in length, but an attacker could supply an arbitrary length name. As a result, the application would crash and execute arbitrary code supplied within the string.

The CIL buffer overrun is a perfect example of a proprietary format vulnerability. The file format was unique to Microsoft; as a result it was not intended to be used except to exchange data between Microsoft applications. It contained both text and binary data, and the length of the text was specified in the design of the format, but not enforced by the application. Finally, because the file was used to package clipart for transmission, it could be exchanged from one user to another or downloaded from the Internet. In fact, CIL files could be downloaded in Internet Explorer without prompting.

Similar vulnerabilities in proprietary protocols abound. The Tabular Data Stream format (TDS) is used to communicate with various database server applications including Microsoft SQL Server. Microsoft SQL Server 2000 had a vulnerability in processing authentication requests using this protocol. Because the vulnerability occurred in the packet that made the initial request for authentication, it has been nicknamed the "Hello" bug. The database instance name, which was supposed to be no more than eight characters in length, could in fact be arbitrary length. Supplying an over-length name would cause the server to crash. In some cases, an attacker could use the vulnerability to execute arbitrary code. Once again, Microsoft assumed that only Microsoft applications (specifically, Query Analyzer) would connect to the server using TDS. Because Query Analyzer's user interface limits the length of the instance name to eight characters, it was thought that the user could not supply additional characters.

USING "FUZZING" TO FIND VULNERABILITIES IN FILE FORMATS AND PROTOCOLS

Fuzzing is the technique of corrupting data in files or protocols at random, in the hope that one of the corruptions might cause the application to crash, hang, or expose a buffer overflow vulnerability. Few mainstream techniques exist for doing fuzzing. *Block-based fuzzing,* as is done by Dave Aitel's tool Spike (described in Chapter 3 and available online at *www.immunitysec.com/resources-freesoftware.shtml*) involves reverse engineering a proprietary protocol and organizing it into basic message blocks. Scripts are created that play these blocks back to form a communication stream. The user inserts corrupt data in the form of "spikes" by specifying that some of the blocks should be substituted with garbage, with very long strings, or out of order messages. You find several advantages and disadvantages to the block-based approach and to tools like Spike. Because the user defines the playback script for Spike, it does not require automation to drive a client application in order to perform testing. Tests are also generally reproducible because the script generates Spike strings in a predictable manner. However, the user must reverse engineer the protocol to create the script; this might not work if the correct values for all fields of a protocol are known, or if there are interdependencies such as lengths or checksums.

The alternative to block-based fuzzing is *proxy-based fuzzing*. A piece of proxy software between the client and server is used to perform corruption on the traffic. This has the advantage of allowing the user to use the client software normally in an attempt to generate traffic, which does not require reverse engineering of the protocol. This technique also works when challenge/response tokens and lengths are in use, provided the proxy doesn't inadvertently modify these. The disadvantage to this approach is that it is somewhat more difficult to corrupt individual fields within a protocol, because the proxy must be programmed to recognize these.

Building a proxy-based fuzzer is relatively easy; we have provided a sample tool, SimpleFuzz, on the CD-ROM accompanying this book in the Source Examples\Chapter 9 folder. SimpleFuzz is written in Visual Basic and can perform three different kinds of corruption. *Random bytes* corruption modifies bytes at random within the communication stream, up to a percentage threshold. It does not modify the length of any transmission. Random bytes corruption is good at finding buffer overflows in binary protocols that depend on positions or sentinel characters (such as "\0") to delimit fields. By corrupting some of these, we might find a field that we are able to overrun into the next. Likewise, if the lengths of certain fields are calculated based on values within the stream, corrupting these might cause an overflow condition to occur.

Long string insertions, on the other hand, do modify the length of the communication stream. This corruption technique inserts a string of predefined length at a certain position in the protocol. The idea is to supply too many characters in a fixed-length field, in an attempt to cause an overrun. Long string insertions work best with text-based protocols.

Substitution corruption searches for a particular string within the communication stream and substitutes it with a different one. This technique can be used to supply long or corrupt values to particular fields within the protocol. For example, protocols that contain text representations of numeric values, or fixed-length strings such as filenames, sometimes are susceptible to buffer overflows when long strings are applied in them.

SimpleFuzz works by being a connection proxy between the client and the server. The client application connects to SimpleFuzz as though SimpleFuzz were the server it was trying to communicate with. SimpleFuzz then opens a connection to the real server and relays data between the two applications. Depending on settings, data from the client to the server, or from the server to the client, is corrupted using one of the corruption techniques. Figure 9.4 shows SimpleFuzz configured to perform long string insertions.

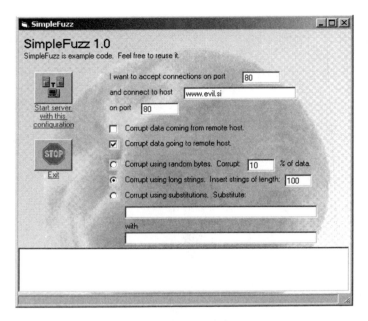

FIGURE 9.4 SimpleFuzz configured for long string insertions.

Suppose we wanted to test a Web application that we thought was susceptible to buffer overflows in a particular CGI field. We can use the substitutions corruption technique to substitute the user-supplied value (entered into the browser) with a long string. This allows us to evade client-side validation of parameters in the browser. So if we wanted to find a buffer overflow in the username field, we might type in a sentinel name—a name that the proxy then substitutes with the long string, into the field. We then configure the proxy to substitute that name with the long string. Figure 9.5 shows SimpleFuzz configured for substitution corruption.

Let's take a closer look at what SimpleFuzz does. It has three main pieces of functionality: the user interface, which just allows the user to set configuration parameters and start the server; the proxying routines; and the corruption routines.

To proxy between the two connections, SimpleFuzz must first set up a server to accept connections from the client. The Start button's click event handler does this:

```
Private Sub btnStart_Click()

' Check the remote server parameters
If tbRemoteHost.Text = "" Or Val(tbRemotePort.Text) <= 0 Then
    lbLog.AddItem ("Remote host name or port number invalid.")
    Exit Sub
End If
```

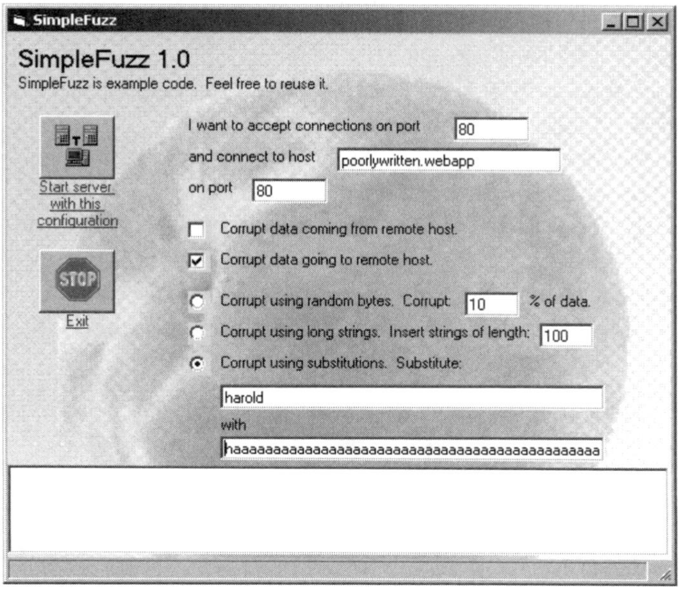

FIGURE 9.5 SimpleFuzz showing configuration for substitution corruption.

```
' Start the server
If Val(tbLocalPort.Text) > 0 Then
    'its a valid port number so connect to it
    tcpServer.LocalPort = Val(tbLocalPort.Text)
    tcpServer.Listen
    lbLog.AddItem ("Listening on port " + tbLocalPort.Text + ".")
Else
    ' its not a valid port number; don't connect
    lbLog.AddItem ("Invalid local port number specified.")
    Exit Sub
End If
End Sub
```

First, we check to see that valid-looking parameters are entered for the local port number, remote host name, and remote port number. The local port number value is used to set the port on which SimpleFuzz listens when the server is started. The other values are used to initiate communication with the real server when a connection is received from the client. We check these now though because there is no sense starting one side of the proxy if the values for the other side are not supplied. Starting a TCP server connection in Visual Basic is easy; setting the port number and telling the Winsock control to "listen." SimpleFuzz uses two separate Winsock controls, one where SimpleFuzz is acting as the client (tcpClient) and one where it is acting as a server (tcpServer).

When the "server" side of SimpleFuzz receives a connection, it must initiate a connection with the real server to start proxying. The code to do this is in the ConnectionRequest event handler for the tcpServer control.

```
Private Sub tcpServer_ConnectionRequest _
(ByVal requestID As Long)
' Check if the control's State is closed.  If not,
' close the connection before accepting the new
' connection.
If tcpServer.State <> sckClosed Then _
tcpServer.Close
' Accept the request with the requestID
' parameter.
tcpServer.Accept requestID
' Connect to the remote server when we receive this request
tcpClient.RemoteHost = tbRemoteHost.Text
tcpClient.RemotePort = Val(tbRemotePort.Text)
tcpClient.Connect
lbLog.AddItem ("Client connected.  Connecting to server.")
End Sub
```

We first check to see if the previous connection to the proxy correctly closed the connection when it exited, because we reuse the same Winsock control for each connection. After setting the control's state to "accept" to complete the handshake with the client, we attempt to contact the real server. We now have two open connections, one to the real client and one to the real server. The proxy exchanges data between them. To do this, we overload the dataArrival event for both Winsock controls. Upon arrival of new data from the client, we corrupt the data and then transmit it to the real server.

```
Private Sub tcpServer_DataArrival _
(ByVal bytesTotal As Long)

Dim strData As String
' get the data from the server
tcpServer.GetData strData

lbLog.AddItem ("Received data: " + strData)
```

```
If cbCorruptFrom.Value = True Then
    strData = Corrupt(strData)
    lbLog.AddItem ("Retransmitted as: " + strData)
Else
    lbLog.AddItem ("Retransmitted intact.")
End If


' send it to the client
tcpClient.SendData strData
End Sub
```

We pass the data to the corrupter only if the "direction" the user has indicated matches the direction this data is traveling in. For the client-to-server direction, this data should be corrupted if the cbCorruptFrom checkbox is checked. (All directions are described from the real client's perspective.) The code to do server-to-client transmission is identical except for direction:

```
Private Sub tcpClient_DataArrival _
(ByVal bytesTotal As Long)

Dim strData As String

'get the data from the client
tcpClient.GetData strData

lbLog.AddItem ("Received data: " + strData)

If cbCorruptTo.Value = True Then
    strData = Corrupt(strData)
    lbLog.AddItem ("Retransmitted as: " + strData)
Else
    lbLog.AddItem ("Retransmitted intact.")
End If

'send it to the server
tcpServer.SendData strData
End Sub
```

The proxy is not configured to close connections on its own. It closes a connection only if one of the two sides has closed its connection to the proxy. When this happens, the proxy must close its connection to the other side. To do this we overload the `close` events of both controls:

```
Private Sub tcpClient_Close()
    tcpServer.Close
    lbLog.AddItem ("Client closed connection.  Closing connection to
server.")
End Sub
```

and

```
Private Sub tcpServer_Close()
    tcpClient.Close
    lbLog.AddItem ("Server closed connection.  Closing connection to
client.")
End Sub
```

Each corruptor is implemented differently, but conforms to a common specification. The corruptors take the data received from the client or server as a string argument and return a string representing the modified data. The random bytes corruption routine is the most complex, because it must decide how many corrupt bytes to insert based on the user-defined threshold. It does this by selecting a random number between 0 and 99 when copying each character. If the number it selected is below the threshold, it substitutes the character it copies with a randomly generated character. This way, the laws of probability are preserved without having to select both a random location and random character.

```
Private Function CorruptRandomBytes(s As String) As String

    ' just in case s is empty
    If s = "" Then
        CorruptRandomBytes = s
        Exit Function
    End If

    ' compute corruption threshold
    cThreshold = Val(tbCorruptPercent.Text)

    ' initialize return string
    t = ""

    For X = 1 To Len(s)

        'initially set to an uncorrupted character
        currentChar = Mid$(s, X, 1)
```

Proprietary Formats and Protocols

```
        ' determine whether to corrupt
        randomNumber = 100 * Rnd(1)
        ' if the random number is less than the corruption threshold
        If (randomNumber < cThreshold) Then
            ' set currentChar to a random character
            currentChar = Chr$(256 * Rnd(1))
        End If

        ' append to return string
        t = t + currentChar


    Next X

    CorruptRandomBytes = t

End Function
```

The long string insertions corruptor uses the substring functions Left$ and Right$ to divide the connection data in half. In between these halves, it inserts a string of "A"s of user-specified length.

```
Private Function InsertLongString(s As String) As String

    ' just in case s is empty
    If s = "" Then
        InsertLongString = s
        Exit Function
    End If

    ' select a random location to insert the string
    inspos = Len(s) * Rnd(1) + 1

    ' set return string
    t = Left$(s, inspos)

    For X = 1 To Val(tbStringLength.Text)
        t = t + "A"
    Next X

    t = t + Right$(s, inspos + 1)

    InsertLongString = t

End Function
```

The substitution corruptor must find a match for its source text within the data stream to perform corruption. It does this by successively comparing substrings of the data stream to the source text. When it finds a match, it replaces that substring with the destination string:

```
Private Function SubstituteString(s As String) As String

    ' if s is shorter than the src string, there's no chance
    ' of a match
    If Len(s) < Len(tbSubstSrc.Text) Then
        SubstituteString = s
        Exit Function
    End If

    ' if either the src or destination string is empty
    ' don't substitute!
    If tbSubstSrc.Text = "" Or tbSubstDest.Text = "" Then
        SubstituteString = s
        Exit Function
    End If

    ' determine whether the src string is in our string
    initialpos = -1
    For X = 1 To Len(s)
        If Mid$(s, X, Len(tbSubstSrc.Text)) = tbSubstSrc.Text Then
            initialpos = X
        End If
    Next X

    ' if it is
    If (initialpos > 0) Then
        ' replace it with the destination string
        t = Left$(s, 1, initialpos) + tbSubstDest.Text + _
            Right$(s, initialpos + Len(tbSubstSrc.Text))
        SubstituteString = t
    Else
        ' otherwise return the original string
        SubstituteString = s
    End If

End Function
```

Substitution corruption in SimpleFuzz has some serious limitations. It cannot substitute more than one time in the same message. It also fails when the source string lies on the packet size boundary—in this instance, parts of the string would be passed to SubstituteString in subsequent calls, but neither would completely match it. It cannot match based on a wild card; a regular expression engine would be needed to do this. However, with some creativity on the part of the user these problems can be avoided.

SimpleFuzz dispatches to one or another of the corruptors based on the values of the radio buttons in the UI. For this reason, a single Corrupt function is called by the data event handlers. This function looks like this:

```
Private Function Corrupt(s As String) As String
    If optRandomBytes.Value = True Then
        t = CorruptRandomBytes(s)
    End If

    If optLongStrings.Value = True Then
        t = InsertLongString(s)
    End If

    If optSubstitute.Value = True Then
        t = SubstituteString(s)
    End If

    Corrupt = t
End Function
```

PREVENTING PROBLEMS WITH PROPRIETARY FORMATS AND PROTOCOLS

The best strategy is to assume all user input is suspect. Never assume that your application is the only one that generates or modifies the data or assume that an attacker cannot modify the data because the format is undocumented. Formats and protocols that are implemented with these assumptions fall down easily when supplied with random data.

A good example can be learned from the Portable Network Graphics (PNG) file format. PNG file chunks each have their own checksum, and the file has an overall checksum that preserves the integrity of the header and individual chunk checksums. PNG also uses a "magic number" technique to detect and correct the most common kinds of corruption. As a result, random data corruptions would have little effect on a PNG file decoder. An attacker would have to have knowledge of the format and correctly specify checksums for each field, but doing this more or less ensures correct data.

Obviously, all data cannot be stored as PNGs. Instead, try to implement the following guidelines when designing a proprietary format or protocol:

- Use a format such as XML, for which (relatively) bug-free parsers already exist. This shifts the responsibility for parsing data to routines that have been designed and tested for completeness.
- Don't rely on length fields or delimiters in protocols to allocate memory. If you must, check that the length is reasonable and meets the maximum constraints of the system including `MAX_FILE_LEN`.
- Never use signed integers when decoding lengths within file formats and protocols. It's too easy for an attacker to use these to find an integer overflow in your parsing code.
- If you're implementing a documented protocol, check a security Web site or mailing list for vulnerabilities in common implementations of the protocol. If other programmers made mistakes in their implementation, you're likely to make the same mistakes. SMTP is the undisputed leader in re-implemented security bugs.
- Remember that the default action for registered file types in Windows is usually to open them in the application that registered the type. As a result, an attacker can trick a user into opening a file containing an exploit even if that file type is not one that is commonly exchanged via e-mail or the Internet, and if the user doesn't understand what application corresponds to the type.
- Remember that if the protocol or format is proprietary to your organization, you are relying solely on your own developers and testers for quality control. If your developers don't know all the security implications of protocol implementation, this is probably best avoided.

Summary Sheet—Proprietary Formats and Protocols

Problem:

Proprietary protocols and file formats rely on security through obscurity and the assumption that attackers cannot modify data they don't understand to prevent security vulnerabilities. This is a false assumption. Even if an attacker does not understand a protocol, unless the decoding routines take active steps to ensure the integrity of data, these routines are commonly susceptible to buffer overflows, format string attacks, and other kinds of parsing vulnerabilities.

Potential Impact:

If an exploitable buffer overflow is found in a file format, any user who opens the file might be susceptible to the exploitation. Common deployment scenarios for these kinds of vulnerabilities include e-mail, Web sites, and file-sharing programs. If a vulnerability is found in a protocol, any application that implements that protocol runs the risk of being exploited remotely.

Even if a remotely exploitable vulnerability is not found, a crash can result in a denial-of-service attack against the server application that implements the protocol.

Habitat:

Hidden, undocumented, proprietary, and special-use protocols and formats are the most likely to be exploited in this manner.

Tools You Need to Find It:

A file corruption utility or network fuzzing tool is used to find these vulnerabilities.

How to Look for It:

Performing corruption on the input data, including random corruption and insertion of long strings, can cause vulnerabilities to manifest.

Symptoms of Failure:

Normal buffer overflow or denial-of-service behavior including crashes in an attached debugger.

Famous Failures/Exploits:

- MS00-15, the CIL buffer overflow vulnerability, is a good example of a proprietary file format bug.
- CAN-2002-1123, the "Hello" vulnerability in Microsoft SQL Server, is a good example of a proprietary protocol that contained an exploitable buffer overrun.

10 Format String Vulnerabilities

In This Chapter

- The Format Family
- Exploiting Format String Vulnerabilities
- Finding This Vulnerability
- References

If you're like most programmers out there, the first program you ever wrote probably output something to a computer screen. In JavaScript, you might choose to pop up a message box with the `alert()` function, or in Visual Basic you might choose `MsgBox()`. Both of these do pretty much the same thing: print whatever string is passed to them in a little pop-up box. If your language of choice is C, chances are one of the first functions you used is `printf()`. `printf` is a fairly basic function; it takes data, formats it, and prints it to `stdout`. The key word here is *formats*. `printf` is part of a class of functions known as *format functions*. In C, these functions take a variable number of arguments, one of which is called the *format string*, which describes the format of output.

Here's a simple example of `printf` in use:

```
//formatit.exe
#include <stdio.h>

int main(int argc, char *argv[])
{
    printf("%s", argv[1]);
    return 0;
}
```

If we compile this program as `formatit.exe`, we can run it with the following result:

```
>formatit hello
hello
```

In this example, the string in quotes is a *format string* and the *format specifier* `%s` tells the function to read the next argument (in this case the `argv[1]`, the first command-line argument) and print it as a string.

The danger with format functions is that input is often printed without a format specifier. For example, in the preceding code we could omit the `"%s"` specifier, which would change the `printf` statement to:

```
printf(argv[1]);
```

Recompiling this program with this change as `formatit2.exe`:

```
//formatit2.exe
#include <stdio.h>

int main(int argc, char *argv[])
{
    printf(argv[1]);
    return 0;
}
```

then compiling and running this program yields the same result as the original `formatit`:

```
>formatit2 hello
hello
```

The difference is that the second executable is vulnerable to something called a *format string attack*. When input is printed with certain specifiers (or without any specifiers), that input itself can be interpreted as a format string. The result can be disastrous with the most extreme case being the creation of an exploitable buffer overflow (see Chapter 9). Consider the following two results.

```
>formatit hello%s
hello%s

>formatit2 hello%s
hello??
```

The difference is that in the first instance we explicitly told the application to treat `hello%s` as a string, and thus it was printed as entered. In the second case, the application interpreted the user input `hello%s` as the string `hello` followed by the format specifier `%s`. When compiled, pointers to the parameters that will be formatted by the specifiers in the format string are placed on the stack. Because in the second case no valid address to a string on the stack exists, the `%s` specified essentially printed (as text) whatever string that was pointed to by whatever memory address was at the top of the stack.

This information might be of little value to an attacker. What would be more helpful is for an attacker to actually read the data stored on the stack. This can be accomplished using the `%x` specifier, which prints (in hex) the 32-bit address at the top of the stack. For example, when `printf` is called, if the stack looks like:

```
a8 44 f9 77
c2 44 f9 77
10 fe 12 00
.
.
.
```

we would get the following result:

```
>formatit2 hello%x%x
hello77f944a877f944c2
```

Note that the values are in little endian order (least significant byte first). Using multiple `%x` specifiers we can look at the contents of the stack. This is a relatively simple attack to carry out, and the result can be the exposure of sensitive data in memory including passwords, encryption keys, etc. Additionally, it is fairly easy to crash the application and cause a denial of service by eventually reading protected memory space or an invalid address (a few `%s` specifiers can usually do the trick).

Several other specifiers perform similar functions. Table 10.1 presents a list of the most important and widely supported ones.

TABLE 10.1 Common Specifiers for Format Functions

Specifier	Description
%d	Signed decimal string (int)
%u	Unsigned decimal string (unsigned int)
%i	Signed decimal string
%o	Unsigned octal string
%x or %X	Unsigned hexadecimal string
%c	Convert integer to the Unicode character it represents
%s	No conversion; just insert string
%f	Signed decimal string of the form xx.yyy
%e or %E	Convert floating-point number to scientific notation
%p	Formats a pointer to an address
%n	Number of bytes written so far
%%	Just inserts %

Format string attacks typically make use of the %x and %n specifiers, although the others can be used to either crash the machine or advance through the stack. Aside from %x, %n is one of the most interesting specifiers because it actually writes something to memory. If we use %n without passing a variable, the application attempts to write a value—the number of bytes formatted by the format function—to the memory address stored at the top of the stack. It is this ability that can ultimately allow an attacker to execute arbitrary commands by taking control of the application's execution path (see the "Exploiting Format String Vulnerabilities" section later in this chapter).

Take a look at how this works. Running under the ntsd debugger with the command line:

```
ntsd formatit %x_%x_%x
```

yields the result shown in Figure 10.1.

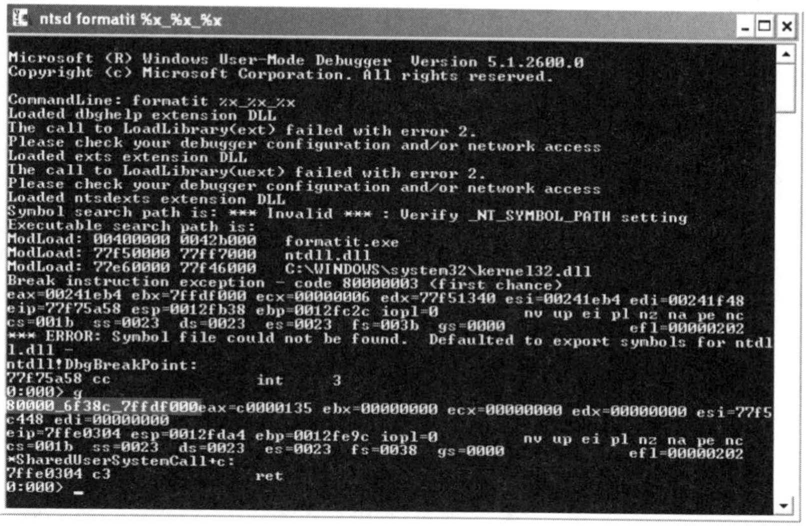

Format String Vulnerabilities

FIGURE 10.1 This figure shows the result of reading three parameters from the stack using the %x specifier. The third parameter, 7ffdf000, we will target to overwrite its contents.

We can see by the output that the third address on the stack is 7ffdf000. Our goal now is to write something to that address, namely the length of our string. The result should be the length of our formatted string, as illustrated in Figure 10.2.

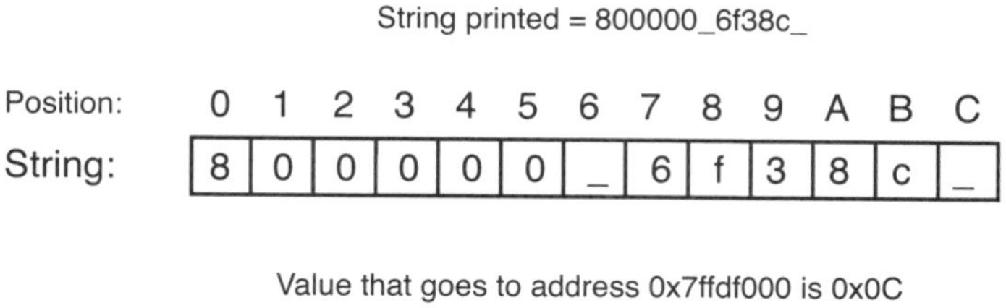

String printed = 800000_6f38c_

Position:	0	1	2	3	4	5	6	7	8	9	A	B	C
String:	8	0	0	0	0	0	_	6	f	3	8	c	_

Value that goes to address 0x7ffdf000 is 0x0C

FIGURE 10.2 The printed string ends at position 13 = 0x0c, and this is the value that is placed at the address 0x7ffdf000.

We can write the length of the string (0x0c) to the address 0x7ffdf000 by using the %n specifier. If we pass %x_%x_%n as our first argument, the result is shown in ntsd in Figure 10.3.

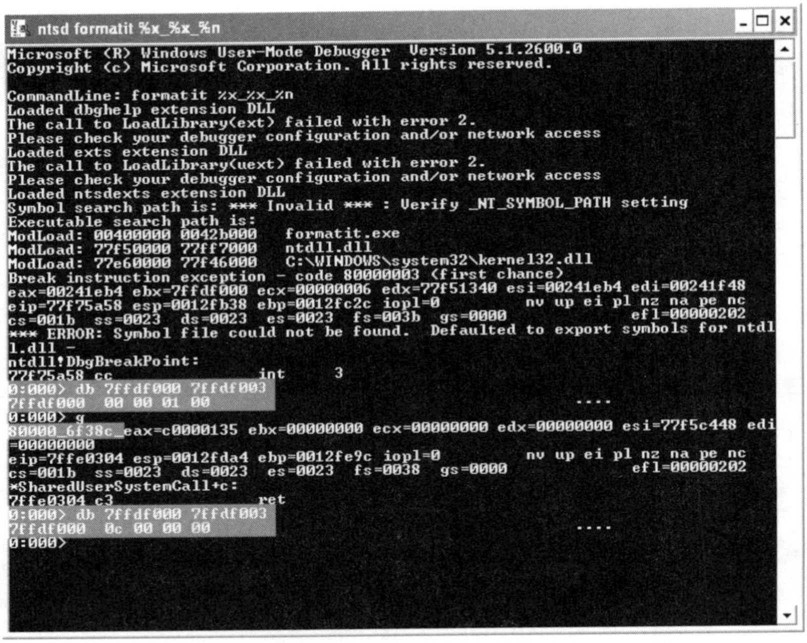

FIGURE 10.3 When we examine the data at address `0x7ffdf000`, we see that it is replaced with the value 0x0c (the length of our string) after `printf()` is called.

Therefore, we can control the value at this address. While this seems like a small accomplishment, an attacker can leverage this ability to overwrite the return address of a function and ultimately execute arbitrary code.

THE FORMAT FAMILY

The `printf` function is a member of a wider class of functions that use format strings for output. Functions such as `sprintf` and `fprintf` are also vulnerable to these types of attack. Table 10.2 lists some other C functions that use format strings and are vulnerable to format string attacks.

Format functions are used to specify the format of output. They can perform conversion so that data types in C are converted into printable form. Besides functions that directly format data, however, a few others such as `syslog` also can process user data and have been exploited through format specifiers.

Format String Vulnerabilities

Of the functions listed in Table 10.2, `sprintf` and `vsprintf` are particularly interesting from a security standpoint because they "print" formatted data to a buffer. Aside from the possibility of a format string vulnerability, using these particular two functions can lead to buffer overflow vulnerabilities also and should usually be replaced with their length-checking cousins `snprintf` and `vsnprintf` (see Chapters 8 and 9 for more information on buffer overflows).

TABLE 10.2 Some Formatting Functions in C Vulnerable to Format String Attacks

Function	Purpose
fprintf	Prints a formatted string to a file
printf	Prints a formatted string to stdout
sprintf	Prints a formatted string to a string buffer
snprintf	Prints a formatted string to a string buffer and the programmer can specify the length of data to be printed to the destination buffer
vfprintf	Prints a formatted string from a v_arg structure to a file
vprintf	Prints a formatted string from a v_arg structure to stdout
vsprintf	Prints a formatted string from a v_arg structure to a string buffer
vsnprintf	Prints a formatted string from a v_arg structure to a string buffer and the programmer can specify the length of data to be printed to the destination buffer

While people have been publicly exploiting buffer overruns since the late 1980s [Seeley89], string format attacks have been well known only since Pascal Bouchareine's 2000 article on BugTraq [Bouchareine00]. However, once the new category was identified, a number of applications were determined to be vulnerable. That year, the Common Vulnerabilities and Exposures database (CVE found at *cve.mitre.org*) lists over 20 major applications and platforms that had been exploited through these attacks, and reports of vulnerabilities continue to flood in up to the present writing.

EXPLOITING FORMAT STRING VULNERABILITIES

We've already seen how an attacker can use %x to read data off the stack, but vulnerabilities from format strings can have much more dire consequences. Consider the code listing that follows:

```
#include "stdafx.h"
#include <stdio.h>


int printstr(char *a)
{
    char buffer[512]="";
    strncpy(buffer,a,500);
    printf(buffer);
    return 0;
}

int main(int argc, char *argv[])
{
    char b[500];
    gets(b);
    printstr(b);
    return 0;
}
```

This code takes in a value from the user and prints it to the screen. While this executable is trivial, the code used here is typical of many large commercial applications. This particular bit of code is wide open to a format string attack because of the vulnerable printf call in the printstr function. Here we see that the application takes a string from the user (using fgets), passes it to printstr, and then prints it using printf. Our first and most basic attack here could be to read information off the stack using a variation of %x, which is %08x. In this case we are telling printf to format the output as 8-digit padded (with 0s) hexadecimal numbers. The result is shown in Figure 10.4.

We can take the output and paste it (as hex values) into a hex editor to get a better idea of what the stack looks like (Figure 10.5).

In Figure 10.5 we see a few memory addresses at the top of the stack are followed by quite a few "c"s and then our data (%08x) in hex. This is valuable information because it tells us how far from the top of the stack our data is located. We now have several options; if our goal was to crash the application, we could simply insert a %s appropriately to read from a bogus memory address causing an access violation. For

Format String Vulnerabilities

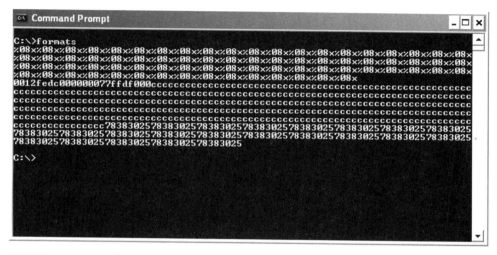

FIGURE 10.4 Reading data from the stack using %x.

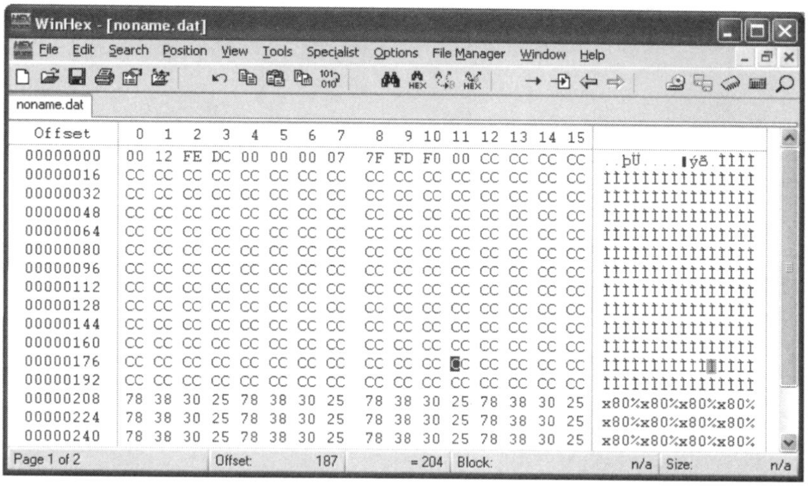

FIGURE 10.5 Putting the output of the data read from the stack into a hex editor gives us a better idea of what the stack looks like.

example, if we wanted to force the application to read from 0xcccccccc, an obviously illegal memory address, we could enter the value %08x%08x%08x%s. The result of doing so is shown in Figure 10.6.

This alone can be an effective denial-of-service attack if delivered to an application remotely. Our ultimate goal here, though, is to execute arbitrary code using

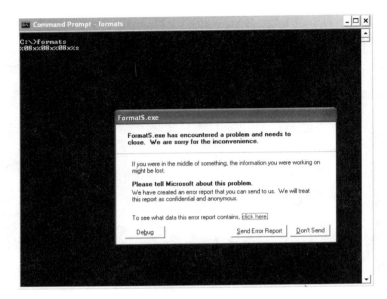

FIGURE 10.6 By forcing the application to read from an uninitialized memory address, we can cause the application to crash.

the format string vulnerability. To do this, we need to be able to write to a memory address whose value eventually gets put into the instruction pointer register (EIP). An obvious target then is to overwrite the return address of the function `printstr`, which is stored somewhere on the stack. We can find out where this address is stored in several ways. The most difficult is to try to "guess" its location based on feedback we get from reading the stack and knowledge of how our application runs. This can sometimes be a painful process, but several techniques have been developed by the hacking community to help. The reality is, though, that for most targets, attackers have access to a copy of the software they are trying to exploit on someone else's machine. For example, if an attacker were trying to exploit a vulnerability in the Apache Web server on a remote host, the attacker would likely set up a mock environment themselves running Apache so that they could study it and use that knowledge to exploit the same software on a different machine. Then the use of a debugger or disassembler to find the likely location of the return address of a function is a reasonable assumption to make in many cases.

Using OllyDbg on our application we can see that the return address to main is stored at `0x0012FC0C` (Figure 10.7).

Format String Vulnerabilities

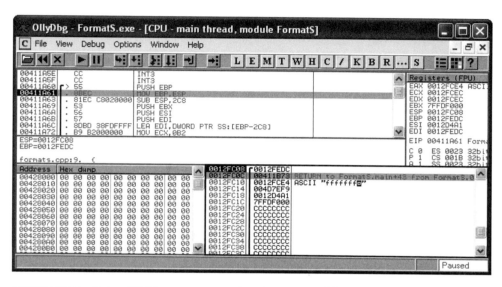

FIGURE 10.7 Using OllyDbg we can see that the return address to `main` is stored at `0x0012FC0C`.

Now, we know what our target address is to write to. We saw in Figure 10.5 that if we continue to walk the stack, eventually we get to the data we originally entered that is stored there. Our strategy is to change the value of the return address so that it points to the beginning of our data in memory. Thus, instead of beginning our string with %x, we enter machine code to be executed by the application. To do this we need a few things:

- A space at the beginning of our string that is eventually replaced with instructions
- The ability to write to arbitrary addresses in memory
- The ability to write arbitrary values to those addresses

Quite a bit of shellcode can be written in 32 bytes. Under the right circumstances, one could, for example, execute an arbitrary application on a machine in just 20 bytes. We therefore use 32 "a"s as a placeholder for our instructions (to be filled in later) at the beginning of our string. The next task is to write to an arbitrary memory location. This can be done by making the address we want to write to a part of our input string and then working our way down the stack until we get to it.

We know the address we want to write to; it is the one we found earlier that holds the return address to `main`, `0x0012FC0C`. The problem is that this address contains a NULL byte, 0x00. Null bytes are usually used to signal the end of a string; therefore, we can't feed the application a null byte through `stdin`. What we can do, though, is take advantage of the fact that the function that reads our data into memory—in this case `fgets`—puts a null byte at the end of our string for us. This means that our target address in memory has to be at the end of our input string, and we have to end our string with the hex values `0c fc 12`. We now run into another problem. The value "0c" represents the ASCII character "backspace." If we try to enter it through the command line with its control character (CTRL-M), the OS interprets it as a backspace, and it deletes the previous character. We can get around this problem by piping the data to the application through a file using:

```
C:>formats < infile.dat
```

where `infile.dat` is the file containing our input values.

Recall before that we used `%08x` to read 4 bytes from the stack. The problem with using this specifier is that we place 4 bytes (the 4 characters `%`, `0`, `8`, and `x`) on the stack, too. Using this method we can never get to the end of our string. If we go back to our original `%x`, however, we get a 2 for 1 move: for every 2 bytes (`%` and `x`) we put on the stack we can advance 4 bytes (the 4 bytes read with `%n`). Our first attempt at an input file then is shown in Figure 10.8.

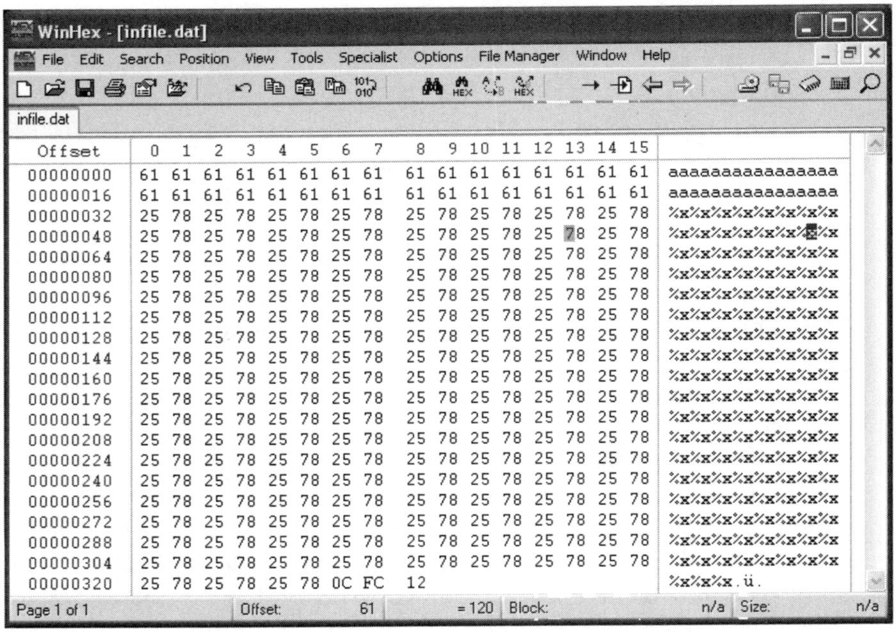

FIGURE 10.8 We make a first attempt at our input file to get to the end of our data on the stack.

Figure 10.9 shows the result of executing our application with this data.

FIGURE 10.9 On our first attempt to get to the end of our input string we overshoot it by several bytes, as can be seen by the trailing 0s.

By removing a few of the %x specifiers from our file we get the last %x to read the value at the end of our string from memory (Figure 10.10).

FIGURE 10.10 Now that we have removed a few of the %x specifiers, we are exactly where we want to be: the last %x is pointing to our data.

We can now replace the last %x with a %n. This causes the value of the number of bytes written by the printf function to be stored at the address 0x0012FC0C: the return address of the printstr function. The modified input file is shown in Figure 10.11.

164 The Software Vulnerability Guide

```
WinHex - [infile.dat]
File  Edit  Search  Position  View  Tools  Specialist  Options  File Manager  Window  Help

infile.dat

Offset     0  1  2  3  4  5  6  7   8  9 10 11 12 13 14 15
00000000  61 61 61 61 61 61 61 61  61 61 61 61 61 61 61 61   aaaaaaaaaaaaaaaa
00000016  61 61 61 61 61 61 61 61  61 61 61 61 61 61 61 61   aaaaaaaaaaaaaaaa
00000032  25 78 25 78 25 78 25 78  25 78 25 78 25 78 25 78   %x%x%x%x%x%x%x%x
00000048  25 78 25 78 25 78 25 78  25 78 25 78 25 78 25 78   %x%x%x%x%x%x%x%x
00000064  25 78 25 78 25 78 25 78  25 78 25 78 25 78 25 78   %x%x%x%x%x%x%x%x
00000080  25 78 25 78 25 78 25 78  25 78 25 78 25 78 25 78   %x%x%x%x%x%x%x%x
00000096  25 78 25 78 25 78 25 78  25 78 25 78 25 78 25 78   %x%x%x%x%x%x%x%x
00000112  25 78 25 78 25 78 25 78  25 78 25 78 25 78 25 78   %x%x%x%x%x%x%x%x
00000128  25 78 25 78 25 78 25 78  25 78 25 78 25 78 25 78   %x%x%x%x%x%x%x%x
00000144  25 78 25 78 25 78 25 78  25 78 25 78 25 78 25 78   %x%x%x%x%x%x%x%x
00000160  25 78 25 78 25 78 25 78  25 78 25 78 25 78 25 78   %x%x%x%x%x%x%x%x
00000176  25 78 25 78 25 78 25 78  25 78 25 78 25 78 25 78   %x%x%x%x%x%x%x%x
00000192  25 78 25 78 25 78 25 78  25 78 25 78 25 78 25 78   %x%x%x%x%x%x%x%x
00000208  25 78 25 78 25 78 25 78  25 78 25 78 25 78 25 78   %x%x%x%x%x%x%x%x
00000224  25 78 25 78 25 78 25 78  25 78 25 78 25 78 25 78   %x%x%x%x%x%x%x%x
00000240  25 78 25 78 25 78 25 78  25 78 25 78 25 78 25 78   %x%x%x%x%x%x%x%x
00000256  25 78 25 78 25 78 25 78  25 78 25 78 25 78 25 78   %x%x%x%x%x%x%x%x
00000272  25 78 25 6E 0C FC 12                               %x%n.ü.

Page 1 of 1           Offset:     275       = 110  Block:            n/a  Size:       n/a
```

FIGURE 10.11 We change the last %x to a %s in the input file, which forces our application to write a value to the memory address 0x0012FC0C, which holds the return address of our printstr function.

When we execute the application with this input file, the application crashes and has the value 0x3df in the instruction pointer (Figure 10.12).

This value, 0x3df, is equivalent to 991 (decimal) which is the number of characters printed by printf. We need to manipulate this value so that it is equal to the address of the beginning of our string in memory. Again we turn to a debugger to find out the address of the start of our string on the stack. Figure 10.13 shows that our string is stored at 0x0012fce4.

If we convert the hex value 0x12fce4 to decimal, we get 1244388. This is the total number of characters we need to be printed. We can force additional blank characters to be printed using the %nx specifier where *n* represents the number of characters to print. For example, the specifier %300x would cause 4 bytes (8 characters) to be printed from the stack preceded by 292 spaces. We are going to use this to our advantage to drastically increase the value of the number written by %n.

Format String Vulnerabilities   165

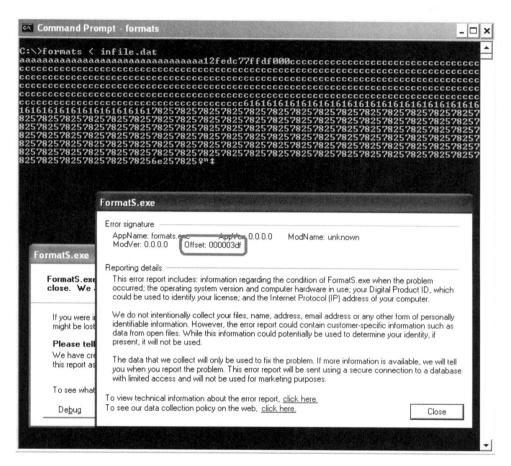

FIGURE 10.12 We have managed to overwrite the return address of our function with the number of bytes printed by format command (0x3df).

The number of bytes we currently need is 1244388 − 991 = 1243397 spaces, which means that the value needs to be 1243397 + 8 = 1243405. Replacing the last %x with %1243405x increases the length of our string by 7 bytes. Because each %x gives us a net move of 2 bytes through the stack, we need to add 4 more %x specifiers and one character to compensate (in this case, we've added the letter "a"). After tweaking the number of spaces slightly to compensate for the added characters, we arrive at the input file shown in Figure 10.13. Notice that we have changed the first several bytes in the file to 0x90. This is a special value in the x86 instruction set, which is interpreted as NOP: no operation. When the application tries to execute this value it simply bypasses it and moves on to the next instruction. Using NOPs is a

166 The Software Vulnerability Guide

powerful technique if we are not sure exactly where in our code the application will land after we overwrite a return address. It is commonly referred to as a "NOP sled" because essentially the application can land anywhere in the sea of NOPs and then "slide" down to our shellcode. Also notice that after the NOPs, we have inserted the hex value 0xCC. This is another special value in the x86 instruction set, which is interpreted as INT 3 (interrupt 3), and when executed, the result is that we break into a debugger.

The value 0xCC is used often by debuggers and represents the INT 3 instruction. When you set a breakpoint in a debugger, what usually happens is that the debugger saves whatever instruction was at the location that you set the break point and replaces it with 0xCC.

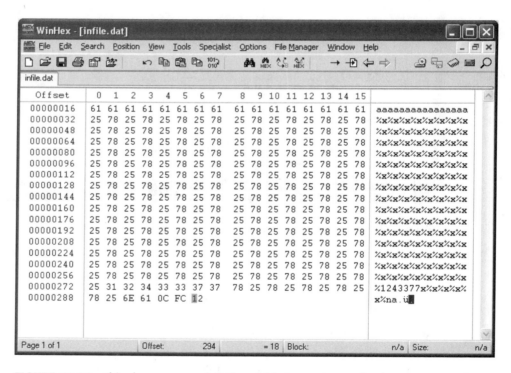

FIGURE 10.13 This shows our exploit file, which forces the application to execute the instruction `CC` stored at the beginning of the file.

After feeding this file into our application, we again get a crash (Figure 10.14). Debugging the application reveals that the reason for the crash was that FormatS was executing our code (in this case CC), which it interpreted as a user breakpoint!

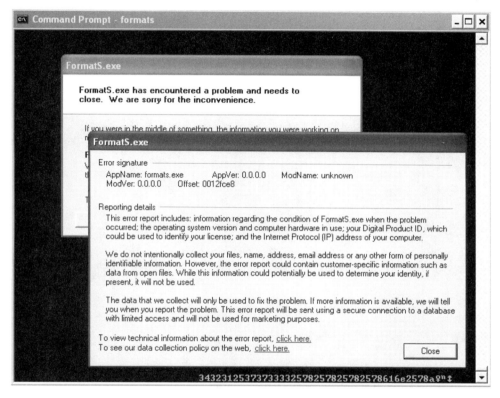

FIGURE 10.14 When we feed the application the data from our modified file, we get a crash at address 0x0012fce8, which is where we have inserted our CC value in memory.

Now we know we can execute code through our data. This is by far the hardest part of the exploit—setting up the injection vector. The only thing that would remain for an attacker is to write shellcode to perform an attack of his choosing. Our goal here, however, is to illustrate the potential severity of this type of vulnerability. These types of bugs must be found and fixed in your software.

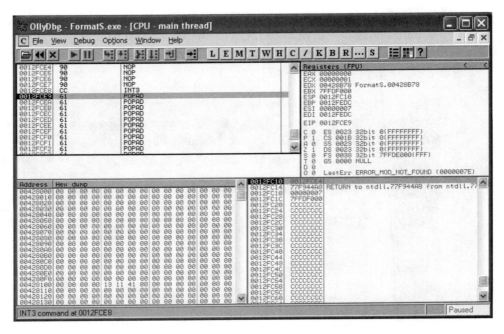

FIGURE 10.15 Indeed, if we then check the crash in a debugger, we see that the application terminated because it executes our breakpoint in memory.

FINDING THIS VULNERABILITY

The good thing about format string vulnerabilities is that they are relatively easy to find in code. Any variable that contains data that is either directly or indirectly influenced by the user should contain a format string that dictates how that data is interpreted. A hand analysis of code can usually find most of the vulnerabilities here. It is important, though, to be familiar with functions that use formatted output. Table 10.2 is a good starting point, but some OS specific functions like syslog() must also be scrutinized.

Also, some automated source-scanning tools for C can make the process of searching through your source code easier. RATS, the Rough Auditing Tool for Security, is a free source code scanner produced by Secure Software (*www.securesw.com*) under active development that is capable of scanning C and C++ source code for format string issues. The ITS4 security scanner by Cigital, Inc. (*www.cigital.com*) is also free and can be used to scan C and C++ for related problems. Flawfinder (*www.dwheeler.com/flawfinder*) is another GPL vulnerability finder that scans C and

C++ code for a variety of security problems including format strings. If your focus is just finding format string vulnerabilities, the pscan tool (*www.striker.ottawa.on.ca/~aland/pscan*) is an open source tool that focuses exclusively on finding format string vulnerabilities in C code.

The biggest problem with using automated scanners is the amount of "noise" or false positives they generate. Practitioners should be sensitive to the fact that not all issues identified by scanners are exploitable. Manual code reviews should look for functions of the form xxprintf() wherever the template is not hard-coded.

From a black-box testing perspective these vulnerabilities can be unearthed by including specifiers such as %x, %s, and %n in input fields. The symptom of failure when using a string of %xs is likely to be "garbage" data returned to the user in a message that quotes the input string. A more drastic approach is to place several %ss into the input string. If a format string vulnerability exists, this causes the application to read from successive addresses at the top of the stack. Because some of the data on the stack is likely to be the contents of other variables (like a string), trying to convert this data to a memory address and then read from that address is likely to result in an "Access Violation" error or core dump, which causes the application to crash.

The key to finding and preventing these vulnerabilities is a keen awareness of the problem.

We encourage you to play with the examples on the CD-ROM that comes with this book to reproduce the specific vulnerabilities discussed in this chapter.

Some of the freely available source scanning tools can identify format string vulnerabilities in applications if you have the source code. These tools include:

RATS: *www.securesw.com/*
ITS4:   *www.cigital.com/its4/*
Flawfinder: *www.dwheeler.com/flawfinder/*
Pscan: *www.striker.ottawa.on.ca/~aland/pscan*

Fixing This Vulnerability

Fixing format string vulnerabilities is easy: use a format specifier to format data. For example, vulnerable calls are likely to look something like this:

```
printf(user_data);
fprintf(stdout, user_data);
snprintf(dest_buffer, size, user_data);
```

If we want user data to be output, processed, or saved, all the preceding functions can be fixed using %s as shown in the following:

```
printf("%s", user_data);
fprintf(stdout, "%s", user_data);
snprintf(dest_buffer, size, "%s", user_data);
```

Summary: Format user data before your users format it for you. To put it even simpler, never let the bad guy supply the template.

Summary Sheet—Format String Vulnerabilities

Problem:

When data from a user is printed with a format function in C or C++, the potential exists that a user can include formatting characters in his data to read memory, crash the application, or execute arbitrary commands. Many common C functions are vulnerable to this attack, including `printf`, `fprintf`, `sprintf`, `snprintf`, and many others. These vulnerabilities can have dire implications but they are relatively easy to find and fix in code.

Potential Impact:

Reading sensitive data from memory, remote DOS resulting from an application crash, and the execution of arbitrary instructions.

Habitat:

Format string vulnerabilities (of the type discussed in this chapter) can be found in applications written in C and C++.

Tools You Need to Find It:

At the source level, scanners like RATS, ITS4, and Flawfinder. Code reviews by good humans.

How to Look for It:

Source scanners such as the ones mentioned previously can be very helpful for finding these vulnerabilities in an automated fashion. For by-hand source code analysis, look for format functions that do not explicitly use a format string to specify how output is to be produced. In most cases these can be fixed fairly easily by adding a format string. For example, instead of printing user data using `printf(user_data)`, use `printf("%s", user_data)`. For black-box testing, try inserting several %s values into strings. If a format string vulnerability exists, this is likely to cause the application to crash.

Symptoms of Failure:

If a user string that is processed by a vulnerable format function contains several %s or %n characters, then the application is likely to crash. Multiple %x characters are likely to result in "garbage" being added to the user string when printed or saved.

Famous Failures/Exploits:

- CVE-2000-0699: Format string vulnerability in ftpd in HP-UX 10.20 allows remote attackers to cause a denial of service or execute arbitrary commands via format strings in the PASS command.
- CVE-2000-0733: Telnetd Telnet server in IRIX 5.2 through 6.1 does not properly clean user-injected format strings, which allows remote attackers to execute arbitrary commands via a long RLD variable in the IAC-SB-TELOPT_ENVIRON request.
- CAN-2004-0354: Multiple format string vulnerabilities in GNU Anubis 3.6.0 through 3.6.2, 3.9.92, and 3.9.93 allow remote attackers to execute arbitrary code via format string specifiers in strings passed to (1) the info function in log.c, (2) the anubis_error function in errs.c, or (3) the ssl_error function in ssl.c.

REFERENCES

[Bouchareine00] Bouchareine, Pascal. "Format bugs." Article appearing in June 2000 issue of *BugTraq*. Online at *www.securityfocus.com/archive/1/70552*. Accessed April 5, 2005.

[Seeley89] Seeley, Donn. "A Tour of the Worm." Available online at *http://securitydigest.org/phage/resource/seely.pdf*. Accessed April 5, 2005.

11 Integer Overflow Vulnerabilities

In This Chapter

- Exploiting Integer Overflow Vulnerabilities
- Finding This Vulnerability
- Fixing This Vulnerability
- References

In a typical buffer overflow, the problem is that developers have allocated a certain amount of space to store data and then the data received is larger than the space allocated. In Chapter 8 we showed how an attacker can leverage this flaw to overwrite arbitrary data in memory and often eventually take control over the application. In this chapter, we deal with an overflow of a different sort: integer overflows. *Integer overflows* occur when we try to store a number in a variable that is larger than that variable's type can handle. For example, depending on the architecture, for the variable declared:

```
unsigned short int num1;
```

most C++ compilers allocate 2 bytes for num1. Table 11.1 shows the typical sizes of integer variables.

For most C and C++ compilers a header file called limits.h defines the size of standard data types.

TABLE 11.1 Names and Ranges of Common Data Types (C and C++)

Type Name	Bytes	Other Names	Range of Values
bool	1	None	false or true
char	1	signed char	−128 to 127
unsigned char	1	None	0 to 255
short	2	short int	
signed short int	−32,768 to 32,767		
unsigned short	2	unsigned short int	0 to 65,535
int	*	signed	
signed int	System dependent		
unsigned int	*	unsigned	System dependent
long	4	long int	
signed long int	−2,147,483,648 to 2,147,483,647		
long long	8	none	−9,223,372,036,854,775,808 to 9,223,372,036,854,775,807
unsigned long	4	unsigned long int	0 to 4,294,967,295

Thus, if a variable is declared as a 2-byte integer, this means that it can take integer values in the range 0 to 65,535. When a value is placed into this variable larger than 65,535, the ISO/IEC 9899:1999 Standard for C compilers states the following:

"The range of nonnegative values of a signed integer type is a subrange of the corresponding unsigned integer type, and the representation of the same value in each type is the same. A computation involving unsigned operands can never overflow, because a result that cannot be represented by the resulting unsigned integer type is reduced modulo the number that is one greater than the largest value that can be represented by the resulting type." [ISO99]

Integer Overflow Vulnerabilities

Translated into English, this means that for an unsigned integer type, an ISO compliant C compiler saves the value of the number to be saved modulus one plus the largest value that type can hold. So, for example, consider the following code:

```
#include "stdio.h"
int main(int argc, char* argv[])
{
    unsigned short int num1 = 65534;
    unsigned int good_num1 = 65534;
    printf("\nSize of num1 \t\t= %d bytes", sizeof(num1));
    printf("\nSize of good_num1 \t= %d bytes", sizeof(good_num1));
    printf("\n\nTrue Value (good_num1)\t\tValue Stored in num1\n");
    for(int i=0;i<10;i++){
        printf("\n%d\t\t\t\t%d", good_num1, num1);
        num1++;
        good_num1++;
    }
    return 0;
}
```

which yields what we see in Figure 11.1.

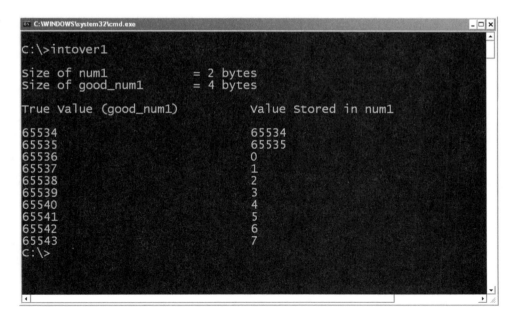

FIGURE 11.1 Overflows in stored values in Standard C.

Here we see that when one tries to store a value in an unsigned integer type that is larger than the maximum value, only the modulus remains, which leads to results like 65535 + 1 = 0. The value essentially "wraps around" to 0. This can cause some unusual behavior if the resultant value is used in a computation. Before we consider some of the consequences, let's see what happens when a signed integer overflows.

For signed integers, the ISO/IEC 9899:1999 standard says that an integer overflow results in behavior "for which this International Standard imposes no requirements" [ISO99]. What this means is that any result is acceptable, including an application crash. Most compilers, however, treat signed integers in much the same way as unsigned integers. Consider the following program:

```c
#include "stdio.h"
int main(int argc, char* argv[])
{
    short int num1 = 32766;
    int good_num1 = 32766;
    printf("\nSize of num1 \t\t= %d bytes", sizeof(num1));
    printf("\nSize of good_num1 \t= %d bytes", sizeof(good_num1));
    printf("\n\nTrue Value (good_num1)\t\tValue Stored in num1\n");
    for(int i=0;i<10;i++){
        printf("\n%d\t\t\t%d", good_num1, num1);
        num1++;
        good_num1++;
    }
    return 0;
}
```

which produces what we see in Figure 11.2.

As with most compiler implementations, here we see a "wraparound effect" where 32767 + 1 = -32766. This conversion occurs not only when the number is incremented, but also when too large a value is stored, as well.

While integer overflows do not allow a user to modify parts of memory directly as many buffer overflows do, the consequences of an integer overflow can be severe. Sometimes the impact of an integer overflow is limited to unusual arithmetic results. In other circumstances, integer overflow vulnerabilities can be leveraged to crash the system or create an exploitable buffer overflow condition.

Consider, for example, the following code listing:

```c
int save_vals(long* array, int number_of_values)
{
    long* savedarray;
    savedarray = malloc(number_of_values * sizeof(long));
```

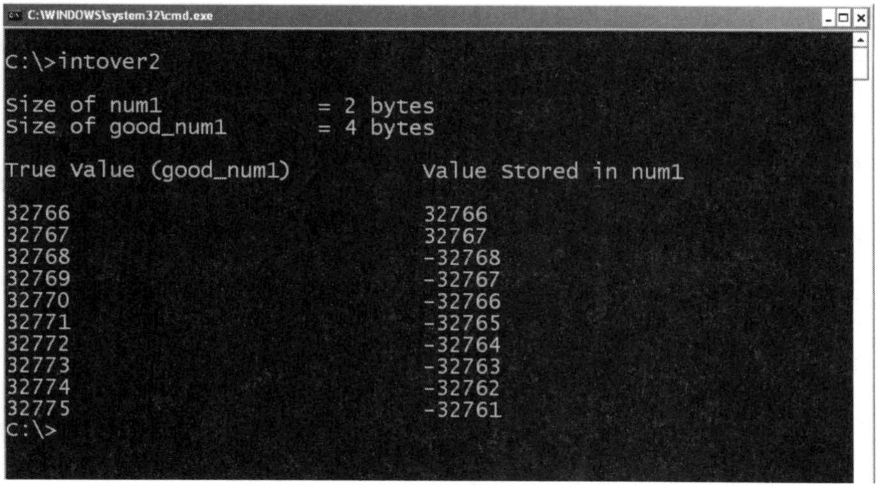

FIGURE 11.2 "Underflow" or "wraparound effect" in signed numbers.

```
        for(int i = 0; i < number_of_values; i++){
            savedarray[i] = array[i];
        }
    output_array(savedarray);
    return 0;
    }

}
```

This function has an insidious integer overflow. Imagine that a user enters a set of values into the main program; these values were saved into an array, and the number of values was recorded. Then both the array and number of values were passed into the function save_vals in the preceding code. If there were only five values in the array, for example, the function would work as intended. Consider, however, if we had a large number of values. On the line:

```
        savedarray = malloc(number_of_values * sizeof(long));
```

a computation is being performed with number_of_values. Assume the size of a long is 4 bytes. This would mean that the result of the computation would be four times the number of values stored. Problems occur, however, when the product is larger than 2,147,483,647, the maximum value of a 4-byte integer. The result is an integer overflow that could be used to force the application to allocate a small amount of space for savedarray and write past its bounds within the for loop. The result is a buffer overflow on the heap that is likely to be exploitable.

Integer overflows can also happen when we perform arithmetic operations on pointers. A pointer is simply an unsigned integer that stores a memory address. Often in the course of writing an application we need to move forward and backward through an array, and sometimes this is done in a loop. If pointer values aren't checked it's easy to make a mistake here and overflow the pointer variable. The results can be unpredictable, but typically the application crashes from a memory access violation.

Beyond pointers, special care must be taken when converting between signed and unsigned integers. The problem here is that it's easy to make the mistake of declaring a variable as an int (allowing both negative and positive values) and then later make the assumption that this variable contains only positive values. We ran into a particularly interesting example of this recently in the Firefox Web browser. When you think about downloading files from the Web, you usually think in kilobytes or megabytes. Ask yourself then, how would you declare a variable that is meant to hold the number of kilobytes remaining in a particular file download? First, you would make it fairly large, and, second, such a variable should contain only positive values and thus should be declared as an unsigned integer. Figure 11.3 shows what happens in the Firefox download manager when we try to download an exceptionally large file. Here we see that the number of kilobytes remaining in a download is negative. We also see that a computation is performed on this value to calculate the download rate in kilobytes per second, also yielding a negative value.

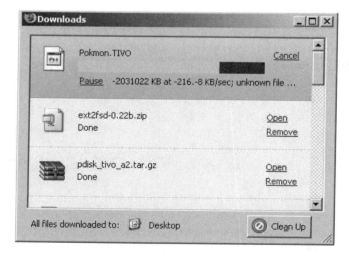

FIGURE 11.3 Download dialog showing an integer underflow.

A closely related issue to integer overflows is *integer underflows*. An underflow occurs when the number we are attempting to store in an integer variable is smaller than the lower bound of that variable type. For unsigned integers, for example, subtracting 1 from an integer with the value 0 stored can result in a very large number. For signed integers, this "wraparound" effect can mean that subtracting 1 from an integer that has its minimal value stored can lead to a positive number as the result.

EXPLOITING INTEGER OVERFLOW VULNERABILITIES

Exploiting integer overflows can be tricky. No cookie cutter method exists that an attacker can use such as the ones used to exploit buffer overflows. The extent of exploitability is very contextual, based on what the errant value produced by the integer overflow is to be used for. Sometimes, it is enough to manipulate a single value to escalate privilege and control the result of some protected or critical value. An example would be an e-commerce application that allows the user to enter the quantity of a particular item that the user wants to buy. If the variable used to store the quantity is of a type whose maximum value is smaller than the quantity entered, the user might be able to make it so that the price charged is a negative amount, which might credit his credit card. Other simple exploits are also possible. You might be able to force a variable to take a value that later results in a failed computation (such as dividing by zero) that crashes the application.

Under the right conditions, integer overflows can create exploitable buffer overflow conditions. For example, if the affected variable or calculation is used to compute the size of data or space allocated for that data, an attacker might be able to manipulate this value to create a discrepancy between the amount of space allocated to store data and the actual size of that data. If an attacker can force an integer that holds the size of input to be stored into a variable, for example, safety checks to make sure that the data to be stored is smaller than the buffer size can be circumvented. In many cases, dynamic allocation of space is done incorrectly, and the result can be a heap overflow. When this happens, an attacker can then use standard buffer overflow exploitation techniques to attack the application.

FINDING THIS VULNERABILITY

Some integer overflow vulnerabilities can be incredibly difficult to find. From a white-box testing perspective, one must not only check integer computations but also the source of values and the context that these values are used in. An integer overflow can occur during an arithmetic operation, a cast, or a copy/read. Few source scanning tools are adept at finding these sorts of problems. The low hanging fruits are casts or conversions from a larger integer type to a smaller one. These are usually brought up as warnings in the compiler. For arithmetic operations that cause an integer overflow to occur, it is often exceedingly difficult to locate the fault in code. You must check all operations performed on an integer-type variable and then determine what the bounds are on the operands. Then, given the maximal (or in some cases minimal) value of the operands, determine if the result could be larger (or smaller) than the resulting integer type.

From a black-box perspective, testing for integer overflows can also be challenging. Ideally, you should reason about the use of all input variables and then speculate as to what calculations might be performed on those variables. Sometimes signs of potential vulnerability appear in error messages returned to the user. Consider, for example, the error message shown in Figure 11.4. This message occurs when we try to insert a table with an obscene number of rows. The result is an error message that tells us the valid range for number of rows is between 1 and 32,767: the maximum value of a short int. This means that the number of rows is likely to be stored in this type of variable. From a testing standpoint, it is interesting to force that variable to take its maximal value and then find an operation that causes it to increment. In the Figure 11.4 example, we could simply right-click on a cell in the table and insert another row. This causes Word to hang on most future operations with that table.

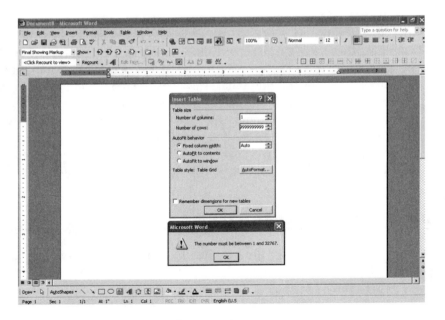

FIGURE 11.4 Duplicating cells in Word.

If we were correct that the variable used to store the number of rows was of type short int, adding an additional row would force the stored value to become −32766, which would lead to some strange computational results as we manipulate the table. The problem with finding integer overflows generally is that the application rarely offers such overt clues. We often have to make assumptions about how a particular input value will be stored and used. The following are a couple specific techniques that have proven useful:

- Try long strings in input fields. While this is the primary technique for locating plain old buffer overflows, it can often be used to find integer overflow issues, too. For example, if the length of the input string is computed, and its value is larger than the integer type it is stored in, the result is an integer overflow. This value can then be used to dynamically allocate memory for the input string and would likely create a buffer overflow condition. When that happens, the result is usually a crash (see Chapter 8 for more information on buffer overflows).
- Try very large (small) values in numeric fields and try to force the application to use these values in a computation. Often, individual values might be appropriately constrained, but calculations might be performed on these values to force the results beyond the appropriate integer range.

FIXING THIS VULNERABILITY

Integer overflows can be fixed by ensuring that the results of computations or the value of inputs are within the appropriate range of the integer type. While the principal is simple, actually constraining values so that they are within the range can be difficult. You must understand the range of operands and then determine the range of the resulting computation. Often this is time consuming. The easiest way to prevent an integer overflow is to test variable values before they are stored or used. Some questions to consider are:

- Should this variable legitimately contain a negative value? If not, declare it as an unsigned int.
- When you compare two integers, are they of the same type (unsigned/signed)?
- When performing arithmetic operations on an integer, can the result be larger or smaller than the maximum and minimum values of that integer type?
- Are sanity checks done before a variable is used in a computation or comparison?
- Are you making trust assumptions about an integer that is being passed into your function (always greater than zero, always positive, etc.)?

Summary Sheet—Integer Overflows

Problem:

Integer data types store values in a specific range. If the application attempts to place a value in a variable of integer type that is larger or smaller than the maximal or minimal value of that type, an "overflow" occurs, and an inaccurate value gets stored. For example, if the value stored in a variable of short int is 32,767 (the maximum value for a short int) and if we increment this value, the variable "wraps around" and the result is −32,766. Aside from inaccurate computations, this might lead to inaccurate length calculations of data and might result in a buffer overflow.

Potential Impact:

Inaccurate computations, application crash, or the creation of a buffer overflow condition that might allow the execution of arbitrary instructions.

Habitat:

Can be found in applications written in a variety of languages but are common in software written in C and C++.

Tools You Need to Find It:

Compiler is good at catching inappropriate integer conversions by giving warnings. Otherwise, finding these vulnerabilities requires hand analysis of the code (white-box) or boundary value testing (black-box).

How to Look for It:

We have two approaches to find integer overflows: code inspection and testing. During code inspection, carefully examine each operand in integer calculation. Determine the range of the operands, the range of the resulting calculations, and the range of the integer variable that is being used to store the computation result. Also, pay particular attention to conversions between integer types and casts. During testing, the goal is to try values at the boundaries of legal values for a particular input field. The goal is to then force the application to use these values in computations, which might precipitate an integer overflow.

Symptoms of Failure:

When an integer overflow occurs, the result is an inappropriate value being used by the application. The failure symptoms can range from overtly wrong calculations to an application crash.

Famous Failures/Exploits:

- **CAN-2003-0357:** Multiple integer overflow vulnerabilities in Ethereal 0.9.11 and earlier allow remote attackers to cause a denial of service and possibly execute arbitrary code via the (1) Mount and (2) PPP dissectors.
- **CAN-2004-0062:** Integer overflow in the `rnd` arithmetic rounding function for various versions of FishCart before 3.1 allows remote attackers to "cause negative totals" via an order with a large quantity.
- **CAN-2004-0431:** Description Integer overflow in Apple QuickTime (`QuickTime.qts`) before 6.5.1 allows attackers to execute arbitrary code via a large "number of entries" field in the sample-to-chunk table data for a .mov movie file, which leads to a heap-based buffer overflow.

REFERENCES

[ISO99] *ISO/IEC9899:1999: Programming Languages – C.* International Organization for Standardization. 1999.

Part IV
Information Disclosure

12 Storing Passwords in Plain Text

In This Chapter

- Finding This Vulnerability
- Fixing This Vulnerability
- References

Many applications and operating systems need to store authentication information to validate username and password data supplied by users. You have various strategies to implement this. Chapter 5 describes Unix passwords in detail. Many of the Microsoft Windows operating systems, for example, store password data in an encrypted file. Aside from being encrypted, the file itself is accessible only to an administrator. Most variants of Linux and Unix use similar strategies. This is arguably a reasonable protection strategy: make password storage accessible only by an administrator and encrypt its contents. Many applications, however, are more frivolous with password storage. When passwords are stored in plain text the possibility exists for an attacker or another user to steal this information. A common error is to store passwords in plain text in authentication files, configuration files, or the Windows registry. Consider the following examples:

Nullsoft Winamp Version 2.8: Winamp is a popular media player that allows a user to play both files locally stored and streaming media. When a remote file is streamed through a Web URL that requires HTTP authentication, the user is prompted to enter a username and password. This username and password is then stored as plain text in the file `winamp.ini` [Bugtraq02]

WinMySQLadmin 1.1: MySQLAdmin, the GUI manager for the Microsoft Windows implementation of the popular MySQL database engine, stores the MySQL password in plain text in the `my.ini` file. The vulnerability exists in version 1.1 of the application. This exposure allows local users to obtain unauthorized access to the MySQL database. [CVE01].

Aside from exposure in files, storing passwords in plain text in the registry can also leave sensitive user data exposed and thus leave the system or specific applications vulnerable, especially if access control lists are not properly used.

During development, care must be taken to adequately protect user authentication information. A common approach is to use a hash function. Implementations such as SHA-1 and MD5 allow developers to store a hash calculated from the user's password. This hash is easy to compute for a particular password, but it is significantly and measurably difficult to calculate the password given the hash. When a user enters his password, a hash is computed based on that password, and it is compared with the stored hash calculated when the password was originally set. Using a hash has the benefit of not having a user's data stored on the machine. Thus, even if a machine storing authentication information is compromised, an attacker would still not have access to authentication information.

FINDING THIS VULNERABILITY

Several strategies to test for plain text passwords exist. One of the most basic is to set a password that is complex and unlikely to occur randomly as a string in other files on the systems. Utilities can then be used to search the contents of files in specified directories for a specific string. For Windows systems, the registry should also be searched.

Finding passwords stored in the registry is relatively easy with the help of an API monitoring tool like Regmon from *Sysinternals.com*. Regmon monitors all calls made to the registry by any application on the system. (Figure 12.1 shows our application's screen.) To do this, we created a small application that prompts a user for a password and looks in the registry for this password. Take a look at what happens when we use Regmon on our demo application.

Storing Passwords in Plain Text   189

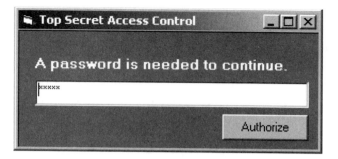

FIGURE 12.1 Application screen.

1. First we add an account with an easily distinguishable password. We chose the password "SECRET."
2. We identify the executable name of our program and launch Regmon. We immediately turn off capturing (using the magnifying glass button) and clear the log (using the eraser button).
3. Regmon allows us to set a filter expression, so we enter the password, "SECRET." Figure 12.2 shows this.

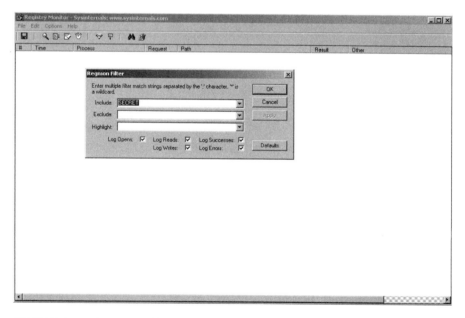

FIGURE 12.2 Regmon allows us to set a filter expression, so we enter the password, "SECRET."

4. We start our application. We enter our candidate password and turn on capturing. Then we click "Authorize."

Figure 12.3 shows the results of our capture. The password was read out of the registry in plain text. We can now browse to this location in regedit, the Windows registry editing tool, and change the password if we want, as shown in Figure 12.4.

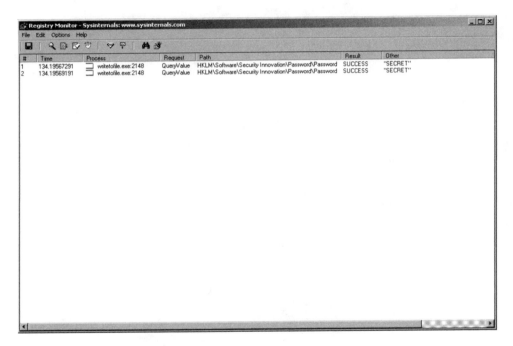

FIGURE 12.3 The results of our capture.

While this kind of password storage might seem foolish, it is quite common. In fact, it is one of the most common vulnerabilities we encounter in non-retail applications. Even some retail applications have had this problem.

A second, more involved approach to discovering password storage is to watch the application execute using a debugger. When a user attempts to authenticate to an application, some comparison must be done between to validate these credentials. In some of the more trivial schemes, the password hash is calculated based on the username. If this is the case, you are likely to find a severe vulnerability that is exploitable through other methods. If, however, a user-supplied password (or some value calculated based on it) is compared with a stored value on the system, you can

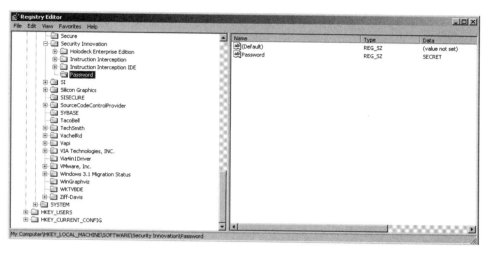

FIGURE 12.4 We can now browse to this location in regedit, the Windows registry editing tool, and change the password.

usually identify the source of this stored data by single-stepping through the application with a debugger once a user has made a login attempt, looking for a "sentinel" password, a known string that "jumps" out in memory. In assembly this usually ends up being a compare statement with a conditional jump afterwards based on the result of the compare. Note that if the password is stored in plain text, the comparison might be character by character.

OllyDbg is an ideal application for this approach because it has a memory search capability. To search the memory of an application with OllyDbg, first load it within the debugger. The application is paused on startup. If you're not sure whether a particular piece of data is in memory at load time, set a breakpoint near where you think the data is used. Note: most constant values (such as hard-coded passwords) are present in memory at load time.

The memory map of the application can be opened by selecting View>Memory or by pressing Alt-M. Right-clicking in this view exposes the Memory context menu. Select "Search." OllyDbg quickly searches the mapped memory associated with an application and finds the first occurrence. Figure 12.5 shows the search dialog box within OllyDbg. Figure 12.6 shows the results of a memory search for the string, "SECRET."

The debugger is also useful in identifying insecure encryption or hashing algorithms associated with passwords. Often times, a password is secured in the registry in a hashed, but easily guessable fashion. We have created a sample application to demonstrate this. This application inputs a username and password and compares it to a hashed password to determine success or failure. Let's look at its implementation.

192 The Software Vulnerability Guide

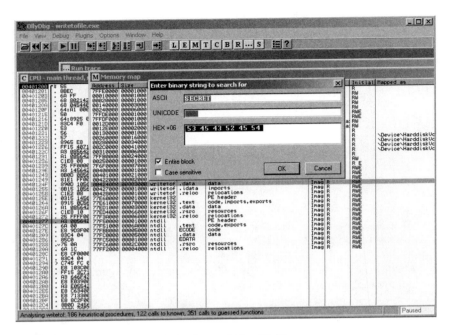

FIGURE 12.5 The search dialog box within OllyDbg.

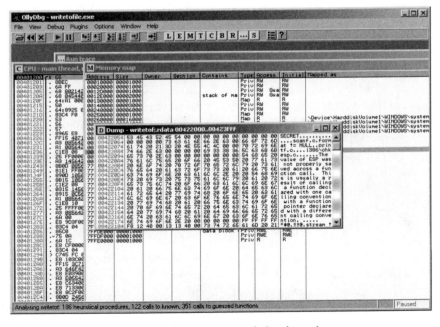

FIGURE 12.6 The results of a memory search for the string.

Storing Passwords in Plain Text

We launch the application in OllyDbg. Right away we can see a few things. First, the text for the username and password prompts is visible and is being passed to an output function. Because we can see these, we can easily identify the area of code where the password input, and potentially computation, takes place. Figure 12.7 shows this.

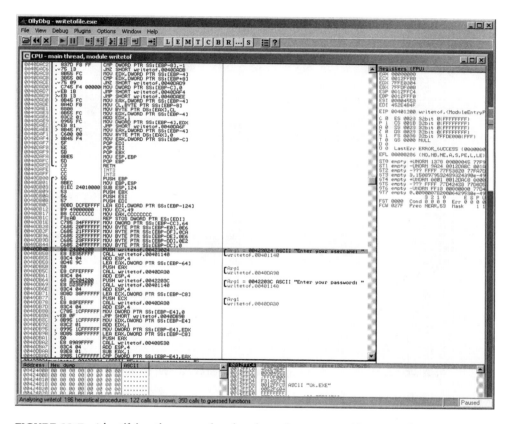

FIGURE 12.7 Identifying the area of code where the password input, and potentially computation, takes place.

We set a breakpoint on the input routine for the username and allow the debugger to run to this breakpoint. From here on out, we single step through the code to see what's going on. After stepping through the inputting of usernames and passwords, we see the values we just entered for the username and password being iterated across in the loop. Each pass through the loop, ECX has one of the characters of the username and EDX one of the characters of the password. In Figure 12.8, which illustrates this, ECX has the value of 6D, the "m" in "simon," and EDX the value 79, the "y" in "says."

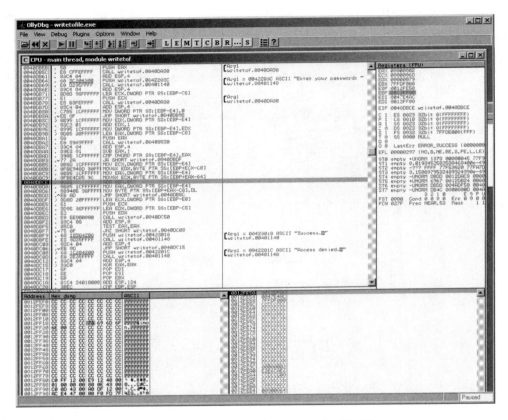

FIGURE 12.8 Looping through username and password in OllyDbg.

Notice what is being done at instruction 040DBCE: they're being added together to form a new string. Letting the loop complete, we follow the code to the comparison with the hashed value. Instruction 040DBED is a call to `strcmp()`; the parameters that have been passed are the new string we just formed and the hashed value from the beginning. The results of this `strcmp()` determine whether the authentication is a success or failure.

Because we've stepped through from start to finish, we can conclude what the algorithm is for password hashing. A hashed password value is formed by adding the bytes of the username and the bytes of the correct password sequentially. This is not strong cryptography and is easy to reverse engineer.

Many applications hash passwords, license keys, and serial numbers using a relatively easy to reverse function. (We would suspect *most* do, but we don't have any data to back that up.) This kind of protection is almost as bad as storing passwords in plain text, because insecure hashes pose no obstacle to the determined attacker.

USING HASH FUNCTIONS

An increasingly popular way to verify passwords is to use hash functions. A *hash function* is a function that takes a relatively arbitrary amount of input and produces a large number. This number (called the *hash value*, or just *hash*) can be extremely useful because it has the following characteristics:

- Most hashing functions are "one way," meaning that if you know the computed hash value, you cannot transform that value back into the original data.
- Small changes in the original data can produce large (and essentially random) changes in the hash value.
- Hash values are fairly evenly distributed across the range of possible values making things like brute forcing attacks very difficult.
- Given the same input data to a hash function, the computed hash value is the same.

Hashing functions are often used to verify that a particular document or character string has not been modified. For example, if we can ensure that the hash value of document A has been kept safe, then we can later compute the hash of document A and compare it with the stored hash value to ensure that the document has not been tampered with.

Hash functions are particularly useful for storing passwords. If instead of storing actual user passwords on a machine, we store hashes computed from the passwords; then even if the machine is compromised, the passwords are not exposed to the attacker. When a user logs in, a hash is computed on the password he entered and is compared with the hash value stored on the system that was originally computed when the password was set.

Several hash functions are widely used; the two most common are the Message Digest version 5 (MD5) and the Secure Hash Algorithm (SHA-1). Both MD5 and SHA-1 are cryptographic hash functions that have the properties outlined previously. Table 12.1 shows the hash values of two strings. As you can see, changing one letter in the string (Text to Taxt) produces radically different hash values.

→

TABLE 12.1 Comparing Hash Values of Slightly Different Strings

	"ThisIsTextThatIsGoingToBeHashed"	"ThisIsTaxtThatIsGoingToBeHashed"
MD5 (128 Bit)	b2647ad639b91b318-fb38b273761e5a7	73fe54f04bf0187-ef2a14719c9a80042
SHA-1 (160 Bit)	b67aeff241d314c29a90a97c8b463453b3c4208d	cf7f47fd5bf9f3d6dd9d3-f18420568428703b159
SHA-1 (256 Bit)	c024c667120cd5a737f222-d1e69fb01d6a8923bfabad82-a93f03d093e592d0ae	e2d279a955d771cc38609b72aae86d641dc4651ec1a32-cd16da4a14dd1d468d2

By our storing hash values instead of actual user passwords, even if a machine is compromised, an attacker is forced to use brute forcing techniques to compare the hash values of millions of strings with the hashed values before a password can be recovered. While several hacking tools exist to automate this process, tools such as L0phtcrack and John the Ripper, using hash values with significant password complexity requirements has obvious advantages over storing actual user passwords on disk.

FIXING THIS VULNERABILITY

The solution to this problem is simple: don't store passwords in plain text. A good alternative to storing passwords in plain text is to use hash functions (see "Using Hash Functions" sidebar). To use hashes effectively you must first compute and save the hashed value of the password when set. While MD5 uses a 128-bit value, some implementations of SHA can have much larger representations and thus prove more difficult to attack. Storing hashes and then taking to precautions to protect these hashes enables you to preserve user passwords in a more secure form. The tradeoff, of course, is that the original password set by the user is not readily retrievable from the stored values. In practical cases this assumption is made by default; i.e., if a user forgets his password, an administrator cannot tell that user what his password was, but he can reset it.

Hashes alone cannot protect passwords, however. An attacker can easily precompute hashes for common passwords; this is essentially the same as a wordlist for plain text passwords. To fully protect a hashed password, it must be combined with a *salt*, a string that is used to perturb the hashing algorithm slightly to prevent precomputation of hashes, or at least increase its complexity.

For example, say that we included a checksum based on the username in the hashed password string. It's easy to combine the username and password together at authentication time to perform the compare. However, an attacker now has to precompute hashes for all username-password combinations, which is much less practical. Further, if the salt is based on a system parameter that is unobtainable to an attacker, precomputation becomes theoretically impossible.

Using the Unix Password Hashing Functions

Unix has built-in support for storing and comparing hashed versions of passwords. The crypt function uses conventional 56-bit DES encryption to create a 13-byte hash of a password 8 bytes or fewer in length. Crypt uses a two-character salt value, which is used to perturb the DES algorithm. The salt is prepended to the password hash so that the same salt can be used in later password comparisons. To compare two passwords, you call crypt on the candidate password using the salt value from the hashed true password and compare the resulting hashes. If they match, the passwords were the same. The code for this is given as follows:

```
char* candidatePassword;
char* hashedCandidate;
/* Constant hashed password value */
/* (the first two characters, "Lr", are the salt) */
char* truePasswordHash = "Lrc1U9A0kOEyW";
char salt[3];

candidatePassword = getpass("Enter your password:");

salt[0] = truePasswordHash[0];
salt[1] = truePasswordHash[1];
salt[2] = 0;
hashedCandidate = crypt(candidatePassword, salt);

if(!strcmp(hashedCandidate, truePasswordHash))
  printf("Success");
else
  printf("Access denied.");
```

Crypt has a number of limitations that make it a less-than-ideal choice for all password applications. The relatively small key length (56 bits) means it can be cracked with a minimal amount of effort by a determined adversary. However, combining password hashing with a lockout after a number of failed attempts can limit brute-forcing of the password. Most versions of Linux support the stronger MD5 family of hashing functions as part of the OpenSSL library. These functions are not as easy to use as crypt(), nor are they as widely supported.

Using `CryptCreateHash` and `CryptHashData` in Windows

The Windows Cryptographic API supports secure hashing of data using a variety of algorithms, including MAC, MD5, and SHA, and is relatively simple to use. First, it is necessary to acquire a handle to the cryptographic provider that will be used. This is because Windows supports multiple cryptographic provider families. `CryptAcquireContext` is used to do this. In our example, we choose the `RSA_FULL` provider, which supports the MD5 checksum algorithm.

```
HCRYPTPROV theCryptoProvider;
CryptAcquireContext(&theCryptoProvider, 0, 0, PROV_RSA_FULL,
    CRYPT_VERIFY_CONTENT);
```

The function `CryptCreateHash` creates a new hashing object associated with a particular hashing algorithm; this object is used by subsequent calls to functions that actually perform the hashing. To create a hash object that uses the MD5 algorithm, we call:

```
HCRYPTHASH hashingObject;

CryptCreateHash(theCryptoProvider, CALG_MD5, 0, 0, &hashingObject);
```

To actually hash a password, we would call `CryptHashData`. `CryptHashData` takes four arguments: a pointer to the hashing function object, the data to be hashed, the data length, and a set of flags, which is ignored. To hash our sample password, "SECRETPASSWORD," we would call:

```
char* password = "SECRETPASSWORD";
CryptHashData(hashingObject, password, strlen(password), 0);
```

Summary Sheet—Storing Passwords in Plain Text

Problem:

When user authentication information such as passwords is stored in plain text, this information is at high risk of theft. Applications, however, routinely store this information in configuration files, databases, or the Windows registry in unencrypted form. Care must be taken to protect passwords from attacker access. One way to do this is through the use of hash functions. Using these functions, only a one-way computed value is stored based on the password. When a user supplies a password, a hash is computed and compared with the one stored.

Potential Impact:

Escalation of privilege, data disclosure, and the compromise of user, application, domain, and system accounts.

Habitat:

Applications that require authentication or applications that manage or use user credentials to authenticate to another application or system.

Tools You Need to Find It:

Binary and text file search tools, such as grep. Debuggers, especially ones with a memory search feature like ntsd.

How to Look for It:

If passwords are stored in plain text in files, databases, or structures (such as the registry), one method is to set a password that is unlikely to occur as a string in other files and search all files, databases, and structures touched by the application under test for this string. A more involved option is to use a debugger and analyze application instructions and data as the application reads in the user-supplied password. If the password is stored in plain text, it is likely that analysis under a debugger can reveal the source of the data that the application is comparing the user supplied password against.

Symptoms of Failure:

File, database, or structure containing user password information unencrypted. This is a difficult symptom to detect, but one that is likely to be uncovered using the techniques outlined previously.

Famous Failures/Exploits:

- CAN-1999-1322: The installation of 1ArcServe Backup and Inoculan AV client modules for Exchange create a log file, `exchverify.log`, which contains usernames and passwords in plain text.

- CAN-2001-1253: Description Alexis 2.0 and 2.1 in COM2001 InternetPBX stores voicemail passwords in plain text in the `com2001.ini` file, which could allow local users to make long distance calls as other users.

REFERENCES

[Bugtraq02] Bugtraq Vulnerability Database ID # 273257. Available online at *www.securityfocus.com/archive/1/273257/2002-05-14/2002-05-20/2*. Accessed April 3, 2005.

[CVE01] Common Vulnerabilities and Exposures (CVE) ID # 2001-1255. Available online at *www.cve.mitre.org/cgi-bin/cvename.cgi?name=2001-1255*. Accessed April 3, 2005.

13
Creating Temporary Files

In This Chapter

- Finding This Vulnerability
- Fixing This Vulnerability
- References

Many applications write data to the filesystem fairly frequently during execution, and many legitimate reasons exist for doing this. Imagine, for example, sorting a long list of names alphabetically. An application might be limited in the amount of memory it has access to, and it might be more efficient for it to use an algorithm that writes some of the data to the filesystem during execution. The end result, however, is likely to be a single file with the desired data, and those temporary files that are no longer needed are then likely to be deleted. In these circumstances, temporarily storing data to the filesystem is not only acceptable, it might be essential.

Programming for efficiency and ease of use, though, sometimes runs directly counter to *secure* programming practices. On some occasions, exposing data to users through the filesystem is a serious security problem. Imagine, for example, an

Internet company whose business model is to sell encrypted music files over the Internet cheaply. The company has convinced music producers to sell their music at heavily reduced rates because each music file will be cryptographically bound to the purchasing user's player. The implication here is that the purchaser can play the music on one and only one machine and that distributing the files to another user would be useless because only the purchaser's player could decrypt the files during playback. From a development perspective, the team responsible for writing the player could either decide to dynamically decrypt the music file during playback—an option that is likely to cause playback to be choppy due to memory demands—or first decrypt the music file temporarily, play the decrypted file, and then delete it after playback. The second option exposes "sensitive" information—in this case the music file—to an attacker. All an attacker would have to do is capture the decrypted file during playback and then distribute it freely. The result is a temporary file vulnerability—which can be exploited easily through automation—that threatens the entire business model of the company.

Products that support any type of Digital Rights Management (DRM) face the same threats. In these cases, the application has access to data that must be protected even from an administrator. For applications that enforce DRM, the implications of information disclosure to the filesystem are obvious. Many applications, however, maintain user or system data that must be protected from other users. Consider, for example, a vulnerability in the RDISK utility in Microsoft Windows NT. The RDISK utility makes a backup of critical system information that can be used to restore the system state. Among the information stored is a complete enumeration of the Windows registry. In Windows NT, the registry contains configuration information, sensitive user data, application data, and, often, application-level passwords. This data could provide a detailed roadmap for an attacker and thus most registry entries are viewable only by an administrator. The RDISK utility, therefore, can only be used by a user with administrative privileges. Figure 13.1 shows the RDISK user interface.

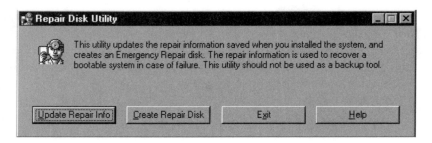

FIGURE 13.1 The RDISK user interface.

Before a recovery disk can be made, the user must first run the "Update Repair Info" utility. This utility extracts information from the registry and saves it to files in the "Repair" subdirectory of the System32 directory. After we run the update utility, we see that several files now exist in this directory. These files contain sensitive system information including registry data and should be accessible only by an administrator. Indeed, if we check the permissions on these files, as shown in Figure 13.2, we see that only an administrator has access to them.

Consider the process, however, that takes place when the repair information is being updated. If we leave the repair folder open and then run "Update Repair Info," we see that a file named $$hive$$ is created and then immediately deleted when the process is done (see Figure 13.3).

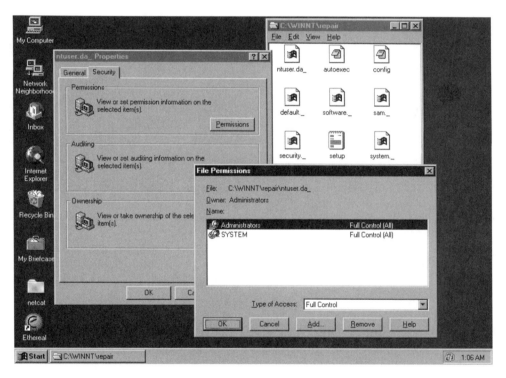

FIGURE 13.2 Checking permissions on sensitive files in the repair folder shows that only administrators have access.

Depending on the memory of the machine and the size of the registry, this process can take anywhere from 10 seconds to several minutes. The file, therefore, exists for only a short period of time and its creation would be noticed only accidentally or by someone explicitly looking for files being created with a utility such

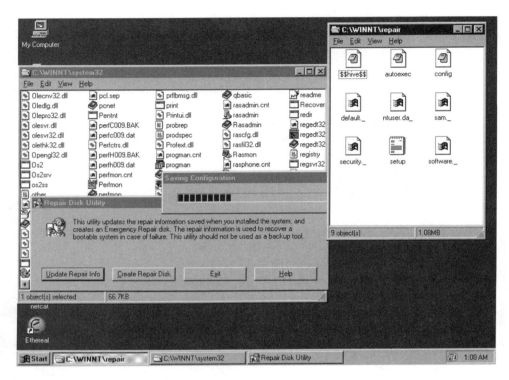

FIGURE 13.3 Watching the `repair` folder while RDISK is backing up information reveals the creation of the temporary file `$$hive$$`.

as Security Innovation's Holodeck (available from *www.securityinnovation.com/holodeck*) or Sysinternals' Filemon (available from *sysinternals.com*). This file contains a complete enumeration of the registry, which would not pose a security risk if permissions on the file were set similar to other files in the folder—administrator access only. Figure 13.4 reveals, however, that permissions on this file are set so that it can be read by any user on the system. This represents a severe security problem. An attacker could create a process that continuously polls for the `$$hive$$` file and then snags it once created. This might require patience in that the attacking process would have to wait until an administrator launched the RDISK utility. As we know though, applications can be very, very patient.

Although data disclosure is the most common consequence of this vulnerability, developers and testers should consider the implications of a user's modifying data contained in a temporary file. For example, a user might be able to take advantage of the fact that a privileged process creates temporary files with improper permissions that contain data that is eventually used by that process. An attacker

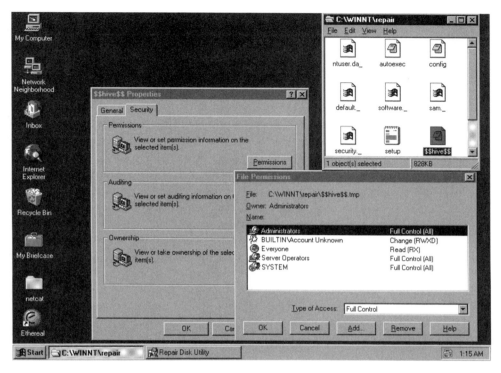

FIGURE 13.4 File permissions for $$hive$$ used by RDISK utility.

might be able to alter that data to either manipulate the behavior of the privileged process or privileged data or cause the application to crash.

These types of failures can be avoided by paying particular attention to the permissions set on files created by your application. Often, we make assumptions about the permissions on files created. Common assumptions are either that the file is created with permissions equivalent to the process's privilege or that temporary files exist for only such a short period of time that we don't care what permissions they are created with. Both are dangerous and both can expose application, user, or system data inappropriately. From a prevention standpoint, developers should ensure that file permissions are set commensurate to the data contained in those files. If *no one* should have access to certain data (as in many DRM applications), then that data should not be exposed to the file system unencrypted.

Overall, temporary file vulnerabilities represent a small portion of the vulnerability types typically discovered in software. They can, however, pose significant data exposure risks and should be considered during both development and testing.

FINDING THIS VULNERABILITY

From a testing point of view, temporary file vulnerabilities can be difficult to find using a strictly UI-based strategy. The biggest issue is that when these types of failures occur during testing, they are often not noticed because interactions between the application under test and the filesystem are not readily visible through the UI. To find these types of problems, we need tools that bring these interactions into plain view during testing. For the Windows platform, we have several options. Security Innovation's Holodeck and Sysinternals' Filemon both allow testers to observe interactions between an application and the filesystem. For Linux and Unix, tools such as strace provide similar capabilities. Identifying file writes is the first step. The next is to investigate their contents and permissions and ask critical questions. Does this file contain information that should be protected from certain system users? Are the permissions on this file commensurate with the users who should be allowed access to this data? Answering these questions might require pausing an application or terminating it with a debugger. For many applications, this can be done using a debugger.

FAMOUS FAILURES: TEMPORARY FILES IN PINE 4.3

Pine, the popular e-mail program for Linux and Unix, had an interesting temporary file vulnerability in version 4.3. If a user configured pine in the following way [SecurityFocus01] (a popular configuration given the widespread use of the vi editor):

```
[x]   enable-alternate-editor-cmd
[x]   enable-alternate-editor-implicitly
editor = /usr/bin/vi
```

pine would create a file in the /tmp directory with the name pico.*pid* where *pid* is the process id of pine. The file contains the contents of the currently edited e-mail.

Several vulnerabilities were found here, all related to using the symbolic link (symlink) operating system feature that essentially creates a placeholder "shortcut" to a file that can be accessed both through the filesystem and programmatically. First, a user could replace one file with a symbolic link to another file so that an attempt to open the original file actually opens the file pointed to by the link. With pine, an attacker could leverage this in two ways. Because by default all users have write permissions to the /tmp directory, an

→

attacker could attempt to predict the process id of pine being run by another user. Take a specific example. Attacker A is targeting user B, and user B is running pine with a process id of 1234. Attacker A would create a link in place of file /tmp/pine.1234 and point it to a file of his choosing. The attacker could have several objectives. The first is to overwrite some file on the system that B has access to and A does not. For example, if victim B has root access, attacker A could then create the symbolic link appropriately and overwrite any file on the system.

A more insidious use of this vulnerability is to create the link to some file created by attacker A that both A and B have write access to. Then, A could hijack B's e-mail message and read or modify its contents.

FIXING THIS VULNERABILITY

Temporary files can be used efficiently and securely. The key is to ensure that sensitive data is protected from users who should not have access to it. The following should be considered before data is exposed in a temporary file:

- Permissions should always be set on temporary files created by the applications to reflect the data contained in those files. These permissions should not allow anyone to access this information who could not access it through other means.
- Particular care must be taken with applications that support Digital Rights Management. In these cases, storing document or media data to a temporary file is not acceptable even if the permissions on temporary files are set so that only administrators can read them.
- Error handlers need to be in place so that if a critical failure were to take place in the application, the temporary files would not be left on disk. Precaution should also be taken that the application cannot be suspended indefinitely while an attacker copies the temporary file.

Summary Sheet—Creating Temporary Files

Problem:

Temporary files are often created by applications in the course of getting their work done. Usually this does not present a security problem. If, however, these temporary files contain sensitive application, system, or user data, care must be taken so that these files are accessible only by authorized users. The most common mistake is creating temporary files that contain privileged data with file

→

permissions that allow less privileged users access. For applications that support Digital Rights Management, protected document or protected media data should never be exposed to the filesystem unencrypted because even an administrator is an untrusted user.

Potential Impact:

Exposure of sensitive data. Possible privilege escalation.

Habitat:

Desktop applications, server applications. Specifically, any applications that manage or use sensitive data.

Tools You Need to Find It:

Environmental monitoring tools such as Security Innovation's Holodeck or Sysinternals' Filemon for Windows and strace for Linux/Unix.

How to Look for It:

Temporary file vulnerabilities can be difficult to find by just looking for on-screen symptoms. Using tools that watch an application's interactions with the filesystem is a must. Several tools such as Security Innovation's Holodeck or Sysinternals' Filemon exist for the Windows platform and equivalent tools like strace can be used on Linux/Unix. The key is to use these tools to watch for file writes and then manually examine file contents for sensitive data. Sometimes, inspecting the files contents might require you to pause or stop execution of your application abruptly so that the file itself is not deleted. You have several options here. The most common is to run your application under a debugger. A less elegant (but effective) option is to kill the application process before it has a chance to delete the file. Once file contents have been examined, the next step is to look at the permissions set on the file. Are they commensurate with who should be allowed access to the data? Generally speaking, temporary files should have permissions that are as restrictive as the most sensitive piece of data that is exposed to the file system.

Symptoms of Failure:

Symptoms are subtle. Might require an observation tool (described previously) to look for application file writes.

→

Famous Failures/Exploits:

- **CAN-2003-0841:** The grid option in PeopleSoft 8.42 stores temporary .xls files in guessable directories under the Web document root, which allows remote attackers to steal search results by directly accessing the files via a URL request.

REFERENCES

[SecurityFocus01] Available online at *www.securityfocus.com/archive/1/150150*. Accessed April 3, 2005.

14 Leaving Things in Memory

In This Chapter

- Description
- Fixing This Problem
- Endnote
- References

As programmers we are frequently reminded to initialize variables before they are used. Most compilers now alert programmers when a variable is used before it is initialized; others safely initialize variables to null. However, with the exception of `free` and `delete`, and occasional destructors added to classes, we do not often think of *unititializing* a variable. In the procedural programming paradigm, most variables go out of scope at the end of the function they are declared in; Java's reference-counting garbage collection actually frees the memory associated with these variables when they go out of scope. The cheapness of memory means that we don't have to be too careful with small memory leaks unless they are in a loop that is executed many times. The vulnerabilities in this chapter are the result of failing to uninitialize data structures when it is essential to do so. Security tokens, passwords, cryptographic keys, and handles to files are among those things that need to be cleared after they are used. When we fail to overwrite or otherwise destroy these data structures, we make them available for an attacker to exploit.

DESCRIPTION

High-level languages play a trick on us. We fall into the old trap of misunderstanding the difference between what something *is called* and what it *really is*—the fact that we lose a reference (or all references, for that matter) to an object in memory does not necessarily mean that the data that represented that object is no longer accessible [1]. Take the following example:

```
void f()
{
  char* x;

  x = (char*) malloc(6);
  if(x)
    strcpy(x,"HELLO");
}
```

Any programmer worth his salt can spot the memory leak this snippet causes. He'll also tell you that without some trickery, we have no way to "get back" x now that we've lost our reference to it, because the pointer was allocated on the stack after f() was called, and the stack is very quickly overwritten by the next function call after f(). But how long does "HELLO" live in memory after x is gone? The answer might be a very long time. Because we lost the pointer to "HELLO" before we explicitly freed it, and it was allocated on the application's heap, it lives at least as long as the application does. Even if we had explicitly freed it using free() or delete, it probably would have stayed right where it was in memory. These operations only make the memory available for garbage collection; they do not destroy the contents. When the application dies, it does not overwrite the data on the heap with zeroes. This would be a woefully inefficient task for an application that consumes a large amount of memory. Instead, the physical pages of memory are released to the operating system, which marks them as unused. When it comes time to use those pages again, the OS might overwrite them with zeroes (Windows does this to comply with Department of Defense (DOD) orange book standards), or it might not. In the worst case, "HELLO" stays in memory until another program actually overwrites it. With 512 MB or more of memory these days, that might be a long time.

Consider a simple password check inside an application. The programmer has been careful not to store the password in plain text on the disk and compares the hashed values instead of decrypting the correct password.

```
char* someusername = "nobody";
struct passwd* pwrec;
char* clearpass;
char* cryptedpass;
char* salt[3];
clearpass = getpass("Password:");
pwrec = getpwnam(someusername);
bzero(salt,3);
memcpy(salt, pwrec->pw_passwd, 2);
cryptedpass = crypt(clearpass, salt);
if(!strcmp(cryptedpass, pwrec->pw_passwd + 2))
   printf("Access Granted.');
else printf("Access Denied.");
```

The programmer might think this code is safe, but that is not entirely true. Notice that getpass() returns a *pointer* to the clear text password obtained from the user. Because getpass() does not know when we'll be done using this pointer, it can never explicitly free or overwrite the memory associated with the input obtained by the user.

So how unsafe is it really to leave this piece of data lying around in memory? After all, the password would be visible in memory only *if* the user is authenticated (otherwise, the password lingering in memory would not be the right password). The answer is it depends on whether you trust every application that runs on your machine. Chapter 22 describes some of the techniques Web browsers use to prevent malicious code from one domain from accessing data belonging to another domain; spyware, viruses, and worms are further examples of executable programs that have access.

It's also important to remember that users tend to use the same password over and over. A password for a user's account on the source code tracking system might be the same as his network password, which might be the same as his local machine's Administrator password. As long as *one* of these is plainly visible to an attacker, the first thing that attacker will do is to try the now-discovered password in as many places as possible, hoping to get lucky. Remember, the odds of one known password working on other machines are *significantly* higher than the chances of being able to successfully crack a password, because people often use the same password on multiple machines.

This problem is especially acute in the area of Digital Rights Management. A rights managed application is one that restricts some of the ways a user can use data within the application. For example, a digital movie application might permit a user to play a movie downloaded off the Internet, but not permit saving that movie to the local hard drive. In a situation such as this, the application cannot trust even a user who has full access to the local machine, because a motivated user could "crack" the Digital Rights Management to illegally distribute the movie.

Digital Rights Management relies heavily on *credentials*, which are encrypted descriptions of the rights a user has to the data. These credentials must be heavily protected in memory, because they are the "keys" to the protected content. Because we don't have encrypted video cards (yet!), the movie's video data must be decrypted somehow to be played to the screen. What prevents an attacker from stealing the decryption key and using it to steal the movie? Short of specialized hardware that prevented exposure of the decryption key (as Microsoft is proposing with its Next-Generation Secure Computing Base (NGSCB) initiative), very little.

> **FAMOUS FAILURES: WINDOWS 98 CREDENTIAL CRASHING**
>
> Sometimes passwords and credentials are intentionally left in memory. This was the case with netlogon usernames and passwords in Windows 95 and Windows 98. In the early days of Windows networking, local machine security was less of an issue than it is today. Windows for Workgroups did not implement secure logon and did not do very much to protect applications from accessing another program's memory. So rather than force a user to re-authenticate each time a share was accessed, Windows for Workgroups implemented a plain text cache of usernames and passwords, which would be tried before asking a user to re-enter his credentials. When credential caching was re-implemented for Windows 95, the "legacy" cache was carried forward. As a result, a malicious attacker could find these usernames and passwords in memory. The credential cache was not even cleared when a user logged off.
>
> Microsoft Security Bulletin MS99-052 (available from Microsoft at *www.microsoft.com/technet/security/bulletin/ms99-052.mspx*) describes this vulnerability.

Finding Exposed Data in Memory

Memory protection, introduced to prevent applications from causing other applications or the operating system from crashing, limits our ability to "peek" and "poke" around in other processes' memory. Each process implements its own "virtual" address space that is not the same from process to process. However, because reading another process' memory is essential to the function of debuggers, both Windows and Linux contain workarounds to accommodate debugging. What's more, both platforms provide a debugging interface that can be used by any program—no debugger black magic is required to access the other process' information. In Windows we can prevent a process' memory from being read by using security tokens; in practice a debugger can attach to most any process if it is launched as Administrator, and most Windows users are logged on as Administrator all the time. In Linux, a debugger can attach only to processes owned by the same user as the debugger, unless the user is logged in as "root."

Searching Memory with a Custom Debugger: Windows

Memsrch.cpp is a memory search tool for Windows included on the companion CD-ROM. Memsrch.cpp searches through the entire process space of an application looking for a given string. Memsrch.cpp works because the Windows Debugging API exposes a function, `DebugActiveProcess()`, that allows you to attach to an already running process as a debugger. To attach to a process, we need to know its process ID, which is obtainable via the Task Manager. The MSDN article "Taking a Snapshot and Viewing Processes" [MSDN05] describes how to do this.

The basic design of this program is as follows:

- Identify what process and string we want to search for.
- Attach to the running process and suspend it so that we can inspect its memory contents.
- Loop through all of the mapped pages of memory, looking for the search string.
- Continue debugging the application. (This is necessary in all operating systems besides Windows XP, because it is not possible to detach the debugger. In Windows XP, a routine called `DebugActiveProcessStop` exists that is able to detach the debugger.)

Our program is as follows. The first several lines include header files, set up the command-line arguments, etc.

```
#include "stdafx.h"
#include <windows.h>
#include <stdlib.h>

int main(int argc, char* argv[])
{

    if(argc!=3)
    {
        fprintf(stderr, "%s: Usage: %s process-id string\n", argv[0], argv[0]);
        return 0;
    }
```

```
// argument one is the process id
int pid = atoi(argv[1]);

// throw it away if it's not a number
if(pid == 0)
{
    fprintf(stderr, "%s: process-id must be a number\n", argv[0]);
    return 0;
}
```

Once we have the process ID, we can call `DebugActiveProcess` on it to attach the debugger and suspend it. We then loop until we get an `EXCEPTION_BREAKPOINT` exception, meaning the running process has been intercepted and is suspended in the debugger.

```
int success = DebugActiveProcess(pid);

// abort if we can't successfully debug the process (bad pid or
debugging disabled)
if(!success)
{
    fprintf(stderr, "%s: could not debug process %d\n", argv[0], pid);
    return 0;
}


DEBUG_EVENT DebugEv; // needed for WaitForDebugEvent

while(1) // loop until we get a breakpoint event
{
    WaitForDebugEvent(&DebugEv, INFINITE);


    if (DebugEv.dwDebugEventCode == EXCEPTION_DEBUG_EVENT &&
        DebugEv.u.Exception.ExceptionRecord.ExceptionCode ==
EXCEPTION_BREAKPOINT )

        break;

    // otherwise keep getting events
```

```
      ContinueDebugEvent(DebugEv.dwProcessId,
         DebugEv.dwThreadId, DBG_CONTINUE);

   }
```

This is the main memory search loop. First, we need to call `OpenProcess` to get a handle to the process, which is used to read the memory. We call this with PROCESS_ALL_ACCESS, which gives us all available rights to the process.

```
      // gain access to the process
      HANDLE hProc = OpenProcess(PROCESS_ALL_ACCESS, true, pid);
```

It would be too inefficient to search every possible address in memory. (It's also not possible, because not every address corresponds to physical memory, and because of memory security.) Instead, we search only those pages that are mapped within an application. To do this, we need to first obtain the page size of the operating system, which is a member of the SYSTEM_INFO structure returned by GetSystemInfo.

```
      // obtain the page size, minimum address and maximum address

      SYSTEM_INFO sysinfo;
      GetSystemInfo(&sysinfo);

      // set up the search string

      char* search_string = (char*) malloc(strlen(argv[2]));
      strcpy(search_string, argv[2]);

      // this buffer will hold the string obtained from the remote
      // process
      char* tmp_buffer = (char*) malloc(strlen(search_string));

      // this is a temporary variable that stores the number of bytes
      // read by ReadProcessMemory
      unsigned long bytes_read;
```

The SYSTEM_INFO structure also tells us the minimum and maximum addresses of the application address space. We use these as the bounds of our search. The VirtualQueryEx function returns a MEMORY_BASIC_INFORMATION structure, which tells us information about the mapping of a particular page within an application. One of the members, meminfo.State, is the state of the memory page; it is MEM_COMMIT if the page is mapped (or committed) to the application.

```c
// search the memory space
for(unsigned long address =
    (unsigned long)sysinfo.lpMinimumApplicationAddress;
    address <=
    (unsigned long)sysinfo.lpMaximumApplicationAddress;
    address += (unsigned long) sysinfo.dwPageSize )
{

    MEMORY_BASIC_INFORMATION meminfo;

    int res = VirtualQueryEx(hProc,
        (void*) address,
        &meminfo,
        sizeof(meminfo));

    if(!res)   // security violation
        continue;

    if(meminfo.State != MEM_COMMIT)   // the page isn't mapped
        continue;

    for(unsigned long p = address;
     p < address + sysinfo.dwPageSize - strlen(search_string) ;
     p ++ )
    {

      ReadProcessMemory (
        hProc,
        (void*) p,
        tmp_buffer,
        strlen(search_string),
        &bytes_read);

        if( !memcmp(
            tmp_buffer,
            search_string,
            strlen(search_string)))
        {
            printf("Search string %s found at address %08x\n",
 search_string, p);
        }
    }

}
```

If we have a valid page, we search the page looking for a matching string. We do this by calling `ReadProcessMemory` successively on each location from the beginning to the end of the block, being careful not to go over the block boundary. (We could use more efficient ways to do this, but they make the code much more difficult to understand.) If the block copied by `ReadProcessMemory` matches the search string, we output the match.

Finally, we should resume the process we're debugging once we're done reading memory. Do this using the `ContinueDebugEvent()` function. It might also be possible to detach from the debugged process; however, we don't recommend this as the process might terminate unexpectedly.

```
// reanimate the suspended process

ContinueDebugEvent(DebugEv.dwProcessId,
      DebugEv.dwThreadId, DBG_CONTINUE);

  while(1) // loop until we get a terminate event
{

    WaitForDebugEvent(&DebugEv, INFINITE);

    if (DebugEv.dwDebugEventCode == EXIT_PROCESS_DEBUG_EVENT)
       break;

    // otherwise keep getting events

    ContinueDebugEvent(DebugEv.dwProcessId,
       DebugEv.dwThreadId, DBG_CONTINUE);

}

// discontinue reading the process

CloseHandle(hProc);


return 0;
}
```

Figure 14.1 shows the completed memory search tool.

FIGURE 14.1 Completed memory search tool.

Searching Memory with a Custom Debugger: Linux

The Linux in process debugging features are all implemented through a single function interface: ptrace(). Ptrace takes a command argument that performs the desired debugging action. To attach to a running process, we would use:

    ptrace(PTRACE_ATTACH, theProcessID, NULL, NULL);

(The process ID in Linux can be obtained via ps.) The first argument to ptrace() is the debugger command to execute. The second is always the process ID of the process we're debugging. The remaining two arguments are pointers that can be used to pass in and pass back parameters to the command; in this case we have no arguments, so we pass NULLs. When ptrace() is called with the ATTACH command, a STOP signal is sent to the child. To wait for the child to stop, use the wait() system call:

    wait(theProcessID);

Once the child has stopped, you can read its memory with the PTRACE_PEEKDATA command. This command can read only one 32-bit word of data at a time, so it must be called successively for each piece of data to be read. To read the heap data in the same way as the Windows example, we would successively call:

    result = ptrace(PTRACE_PEEKDATA, theProcessID, address, NULL);

with address set to each address we wanted to read.

Similar to `GetThreadContext()`, `PTRACE_GETREGS` enables you to get the context of the debugged process, including its register values. Likewise, `PTRACE_CONT` resumes the debugged process in the same way `ContinueDebugEvent()` does.

FIXING THIS PROBLEM

The simplest way to cure this problem is to delete security-relevant data in memory as quickly as possible after it's used. The functions `bzero()` in Linux and `ZeroMemory()` in Windows accomplish this. Microsoft recommends using `SecureZeroMemory()` instead of `ZeroMemory()` in secure applications, because compilers might optimize out calls to `ZeroMemory()` without the programmer's knowledge. You can always explicitly `memset()` data structures to zero as well. The real problem lies in system calls that do not securely overwrite memory after it is used, or cannot because of design limitations. Functions like `getpass()` fall into this category. Most folks would assume that `getpass()` is more secure than a function they could write themselves. The contrary is true. Whenever possible, avoid using these functions without carefully reading the documentation and considering your potential application. Don't assume.

Summary Sheet—Leaving Things in Memory

Problem:

Data can persist in memory for long periods of time after it is used. Even explicitly freeing memory using a `free()` or `delete()` function will ensure that it is no longer accessible. Certain data structures, such as passwords, can become the target of attackers through the use of malicious programs running on your machine.

Potential Impact:

An attacker might recover an important piece of confidential data such as a password or encryption key.

Habitat:

Any environment where all of the running programs cannot be trusted is a target.

Tools You Need to Find It:

A debugger that can search memory (OllyDbg is a good one for this) or a custom-written memory search tool.

→

How to Look for It:

Probe the heap area of running applications for the signatures of passwords, encryption keys, etc.

Symptoms of Failure:

Try scanning for a known password and see if you can see it within memory.

Famous Failures/Exploits:

- MS99-052 "Legacy Credential Caching in Windows 95 and Windows 98." The sidebar in this chapter describes this vulnerability in detail.

ENDNOTE

[1] The "White Knight's Song" in Lewis Carroll's *Through the Looking Glass* is about the difference between what something is and what it's called. This song is sometimes used to teach computer science students about pointers. In addition to being a children's author, Carroll taught mathematics and logic at Oxford University. For information about Lewis Carroll, you can visit *http://lewiscarrollsociety.org.uk/*.

REFERENCES

[MSDN05] "Taking a Snapshot and Viewing Processes." *Microsoft Developer Network (MSDN)*. Available online at *http://msdn.microsoft.com/library/default.asp?url=/library/en-us/perfmon/base/taking_a_snapshot_and_viewing_processes.asp*. Accessed April 3, 2005.

15
The Swap File and Incomplete Deletes

In This Chapter

- Using a Disk Editor to Find Confidential Data Fragments
- Fixing This Problem

Both vulnerabilities in this chapter are a result of differences between the *abstract* behavior a programmer expects from the operating system and its *real* behavior. In the simplified view we are taught in programming classes, memory is memory and files are files. Additionally, we tend to think that objects cease to exist if we can no longer reference them. The reality of finite storage (disk and memory) is that it always contains *something*, even if we think of it as containing nothing. The thing it contains might be your application's confidential data—an unencrypted copy of a now-encrypted file, the password to a network share or database, or cryptographic key material. An attacker can use techniques to recover this material if careful steps are not taken to prevent it from being exposed.

Our first vulnerability deals with a feature that has been present in a multitude of operating systems dating back to the early days of computing. Virtual memory (sometimes called a virtual store, paging file, or swap file) allows applications to make use of more memory than is physically present in the system by copying data in the physical memory into secondary storage (such as a hard disk) and reassigning that physical memory to another application. The data can be moved back into a new piece of physical memory later when it is needed. The OS accomplishes this by means of a *memory map*, a table that the CPU or memory management unit (MMU) uses to translate the virtual addresses applications use into physical addresses. The virtual memory manager software in the operating system interacts with the MMU to do the swapping to and from disk in a way that is transparent to applications.

As a result, data that looks like it is only resident in memory from an application's point of view might have actually spent some time on the disk. This is dangerous if this information is something you are trying to prevent a user from accessing, such as an encryption key or password. To make matters worse, because the OS must move data in and out of memory as quickly as possible, it does not delete the remnants of your data on the disk after it is swapped back into memory. Instead, it remains until that page in the page file is overwritten, or the page file is cleared and recreated.

A malicious user with access to the local machine can use "efficiency" of the operating system to steal the confidential data directly out of the page file. The real risk of this depends significantly on the application and the environment in which the software is running. In most cases an attacker would have to have direct physical access to the machine in order to read the page file, and might have to boot it using a boot image from a different operating system; i.e., boot a Windows system with a Linux CD. As a result, this is not a security risk for ordinary applications. However, some applications have secrets well worth hiding even from a user with physical access.

Consider an application that uses a password and salt to create a symmetric encryption key that is used to protect data stored on the disk. We could use the OpenSSL Blowfish algorithm to do this:

```c
#include <openssl/blowfish.h>
#include <stdio.h>

#define DATALEN 8

void main(int argc, char* argv[])
{
  FILE* infile;
  FILE* outfile;
```

```
    char in[DATALEN+1];
    char out[DATALEN+1];
    char* password;
    BF_KEY key;

    infile = fopen(argv[1],"r");
    fread(in,DATALEN,1,infile);
    fclose(infile);

    password = getpass("Password:");

    BF_set_key(&key,DATALEN,password);
    BF_ecb_encrypt(in,out,&key,1);
    bzero(in,DATALEN);
    bzero(key,sizeof(BF_KEY));
    outfile = fopen(argv[2],"w");
    fwrite(out,DATALEN,1,outfile);
    fclose(outfile);
}
```

This example was selected in order to illustrate a vulnerability; Blowfish is generally not considered secure unless it has a 128-bit key.

The structure key points to the result of the set_key() function, which contains the key used to encrypt the data. What happens if the portion of this application's memory that contains this array is swapped out in the middle of execution? The answer is that the key is written out to the page file, where an attacker could find it and use it to decrypt the data. While this might require physical access to the machine (the page file can usually be read only when the machine is booted with an alternate boot device), in some circumstances this is still a concern; for example, sensitive national security information.

It's also possible that the page containing in could be swapped out between calling fread() and calling bzero(); this would mean that though the unencrypted text is destroyed in memory, the original unencrypted data would be left behind in the page file.

A related vulnerability exists because of the way operating systems read, write, and delete files. In most disk systems, data can be randomly accessed only in fixed sized blocks. This is at odds with the idea of arbitrary-length files, which are a fixture of most applications. To disguise this to both the user and applications, the operating system divides a file into several blocks of the appropriate size and stores them on the disk, keeping track of their location using an indexing system or set of pointers. Files are most efficiently read and written by the disk hardware when they are *contiguous*; that is, all the blocks that make up a file follow one after another on the disk. So when a new file is created, the OS tries first to find a sequence of blocks large enough to fit the whole file contiguously before reusing empty blocks on the disk. This has two consequences. One is *fragmentation*, a performance issue that arises from not having enough large sequences of blocks to go around, resulting in each large file being split up and spread over the disk. The other consequence is one of insecurity and might allow an attacker to recover portions of confidential data even after it is deleted.

We mentioned that the OS wants to store files in contiguous blocks. What happens if you delete a small file? While the OS marks the file as deleted in its index (or perhaps just removes the reference to it), it does not necessarily overwrite the blocks that once contained that file. Consider the Blowfish example we gave previously. What will the user do with the file pointed to by `infile` that contains the unencrypted data? If he's smart, he'll delete it; there isn't much point in encrypting something and leaving the unencrypted copy around to be read. But whether the file is deleted by the user through the operating system, or the programmer through the encryption application, in all likelihood an OS function such as Unix's `remove()` or Windows' `DeleteFile()` will be used. These functions do the unlinking from the index, but do not overwrite the data in a secure manner. Unlike the page file vulnerability, in which there's just a small chance the key or unencrypted data is left behind where someone can access it, the data from the deleted file *almost certainly* remains on the disk, at least for a while. In fact, the undelete feature of operating systems relies on this to recover a user's accidentally deleted file with a high chance of success.

USING A DISK EDITOR TO FIND CONFIDENTIAL DATA FRAGMENTS

Raw disk editors are special hex editors that are designed with knowledge of the filesystem's format and layout. Norton's Disk Doctor, which has been around since the earliest days of the IBM PC, is one of these. Some ordinary hex editors, including WinHex, have this feature built into them. The standard utility `debugfs` can be used to edit ext2 filesystems under Linux.

Most tools (and most OSes) don't allow raw editing of disks while they're in use by the operating system, so it is handy to have another means of booting the computer besides the main OS partition. Once this is accomplished, make a copy of what you want to edit, even if you don't plan to change anything. Raw editing of disks is inherently dangerous to the data on them, especially if you're not careful.

Most of the safeguards against accidental deletion of important data (like the file indexes) are turned off in these programs.

In Windows NT, 2000, and XP, the page file is called `pagefile.sys` and is usually located in the root directory of the main partition, though the user can change the partition he wants to use. It is a hidden file; use the command: `dir /a:h` to find it. In Linux, the operating system swaps to a specially made partition. Look in `fdisk` for a partition with the ID number set to 82 hex. Once you have identified it, you can use the `dd` command to copy it to a file. For example, if your swap partition was located on the device `/dev/hda2`, the command would be:

```
dd if=/dev/hda2 of=tmpfile
```

The contents of the swap partition would be copied to the file `tmpfile`.

Finding deleted files is a little bit more tricky, though some disk editors can do this. If you don't have one of these, you might have to search the whole partition, which can take some time given the large size of today's hard drives. One of the best ways to access a whole partition in raw mode is to boot with a Linux live CD—Knoppix from *Knoppix.net* is a good source for this. Live CDs allow you to run the entire OS from off a CD without touching any of the hard drives installed in the machine. They typically creates a RAMDISK scratch area where you can edit files, etc., but this data is lost upon reboot. Once you have booted with the live CD, you can access the hard drives in raw mode. To back up the main partition (usually `hda1`) you can mount a remote share and copy the data over the network:

```
mount -t nfs myserver:/myshare /mnt
dd if=/dev/hda1 of=/mnt/tmpfile
```

Finding deleted data on a disk is relatively easy if you know what you're looking for. We created a simple text file containing the word NORMANDY. Using the Blowfish program, we encrypted this file and saved the result as ENCRYPTED.TXT. After encrypting, we deleted the original. Figure 15.1 shows the floppy drive contents in Explorer showing only the ENCRYPTED.TXT file.

WinHex supports direct disk editing. We used the Disk Editor feature under Tools and selected "00h Floppy Disk 1." The raw contents of the floppy are brought up in the editor window. You can see in Figure 15.2 the skeleton bootloader Windows places on the floppy to tell a user the disk is not bootable.

Scrolling down, we can see the File Allocation Table, the index that maps filenames to blocks on the disk. We can see several filenames that we couldn't see in Windows Explorer: each of these filenames begins with hex character E5. This is how the FAT filesystem marks a file as deleted. Our original file, SECRET.TXT, is there, with the "S" changed to an E5. Notice that even the New Text Document.txt

228 The Software Vulnerability Guide

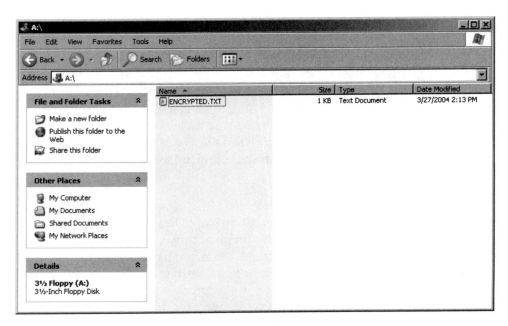

FIGURE 15.1 Explorer showing encrypted text file.

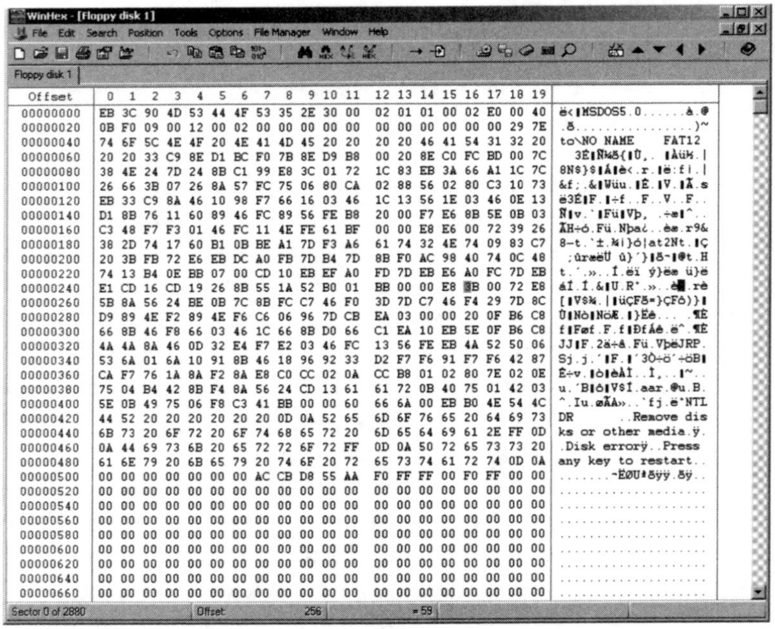

FIGURE 15.2 Disk editor showing skeleton bootloader.

filename, which Windows creates and prompts you to rename when you create a new text document, is still present in the index. Figure 15.3 shows these "ghosts" in the File Allocation Table.

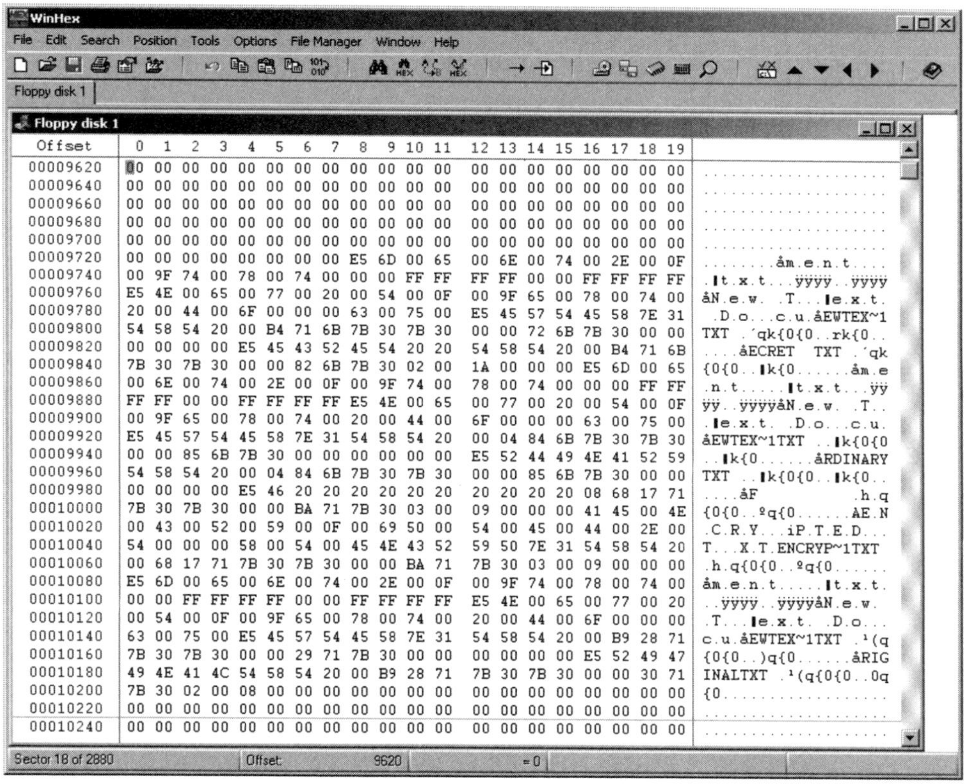

FIGURE 15.3 Ghosts of deleted files in the File Allocation Table.

Knowing the format of the index, we can follow the pointers to our original, unencrypted file. The index formats for commonly used filesystems are easily found on the Internet; even the officially undocumented NTFS filesystem is relatively well-understood. Lo and behold, at location 16880 we find our original, unencrypted text, as illustrated in Figure 15.4.

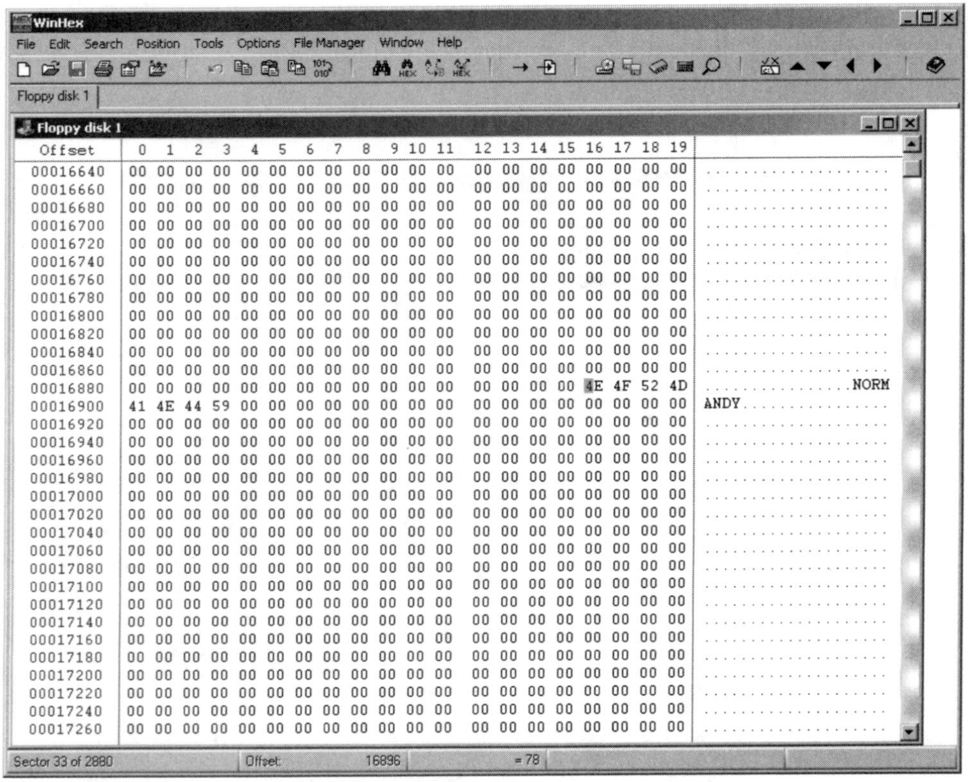

FIGURE 15.4 Original, uncencrypted text found in disk editor.

FIXING THIS PROBLEM

Later versions of Linux and Windows both implement page locking as a way of fixing the page file vulnerability. In Windows, the VirtualLock() API function can be used to protect a page segment from being swapped to the disk. When an address and range are passed to VirtualLock(), any associated pages are not swapped until VirtualUnlock() is called on them or the process terminates. Once a process terminates, the page is marked as free, and the OS attempts to reuse it. However, because Windows zeroes dirty pages before they are given to another process, it is unlikely that the page will be swapped out with the confidential data on it after process termination. VirtualLock() has one limitation: for performance reasons, Windows limits the number of pages that can be protected with VirtualLock(). Therefore, in a high memory usage situation where pages are likely to be swapped, VirtualLock() might be unable to lock a page. Short of using VirtualLock(), an end-user solution would be to cause Windows to delete the page file upon shutdown. To do this, edit the registry key `HKEY_LOCAL_MACHINE\SYSTEM\CurrentControlSet\Control\Session Manager\Memory Management` and add the attribute `ClearPageFileAtShutdown = 1`. This forces Windows to delete the page file *when it is shut down properly*. No user-level protection exists in the event of an improper shutdown.

Linux implements memory locking in a very similar fashion to Windows. The `mlock()` system call is used to lock memory, and the `munlock()` call is used to unlock it. Look at our example code again, this time using `mlock()` to protect the important memory elements from being swapped.

```
#include <sys/mman.h>
#include <openssl/blowfish.h>
#include <stdio.h>

void main(int argc, char* argv[])
{

  FILE* infile;
  FILE* outfile;

  char in[9];
  char out[9];
  char* password;
  BF_KEY key;

  infile = fopen(argv[1],"r");
  mlock(in,8); /* lock the in buffer in memory */
  fread(in,8,1,infile);
  fclose(infile);

  password = getpass("Password:");

  mlock(key,sizeof(BF_KEY)); /* lock the key in memory */
  BF_set_key(&key,8,password);
  BF_ecb_encrypt(in,out,&key,1);

  bzero(in,8);
  bzero(key,sizeof(BF_KEY));

  /* unlock only after we've zeroed all the data
     structures */
  munlock(in,8);
  munlock(key,sizeof(BF_KEY));
```

```
outfile = fopen(argv[2],"w");
fwrite(out,8,1,outfile);
fclose(outfile);
}
```

From a practical point of view, our augmentations might be excessive. Because `in`, `password`, and `key` are all on `main()`'s stack, they almost certainly reside on the same page, so a call to lock any one of them locks all three data structures. Likewise, all pages are unlocked when the application terminates, which it does shortly after the `munlock()` calls. However, in a more complicated program this might not be the case. To be safe, we lock each data structure we don't want written to the disk, and unlock it only after we've zeroed the data structures in memory.

Fixing the deleted file problem is much more difficult. No API is provided for this on either platform. Windows does zero the disk blocks *when they are reallocated to another application*; however, this does nothing to protect the data between deletion and reallocation, which can be a very long time. Because neither Windows nor Linux allows you to access the disk in raw mode without a great deal of dangerous hacking and patching, it might not be possible to do this programmatically. A number of utilities exist to do this, but they are generally GUI-driven and of no use to developers. The safest bet is to avoid writing sensitive data to the disk in an unencrypted fashion whenever possible.

Windows provides an easy mechanism to encrypt data stored on the disk—the `FILE_ATTRIBUTE_ENCRYPTED` attribute, which can be passed to `CreateFile`. An encryption routine for already existing files, `EncryptFile`, and a decryption routine, `DecryptFile`, also exist.

Summary Sheet—The Swap File and Incomplete Deletes

Problem:

The operating system can swap portions of memory containing secure data out to the disk. It does not overwrite this data when the memory is freed or the process owning it is terminated. Additionally, the data contained in a file is not completely removed from the disk when the file is deleted. The same techniques that can be used to recover an accidentally deleted file can recover this data. As a result, confidential data might be stored on the disk without the programmer's knowledge.

Potential Impact:

An attacker might recover an important piece of confidential data such as a password or encryption key.

→

Habitat:

Applications that guard important secrets such as passwords or perform encryption are most at risk. The risk is also elevated in multi-user environments where one user might be able to recover another user's data.

Tools You Need to Find It:

A disk editor or hex editor that is able to do raw disk access.

How to Look for It:

Identify critical sections of an application that have data that should not be swapped to a disk or left behind after deletion. Inspect the swap file and deleted filesystem nodes for evidence of this data.

Symptoms of Failure:

Obvious indications of "ghost" data, such as large blocks of unencrypted text or the signatures of key material data structures.

Famous Failures/Exploits:

- On older versions of Red Hat, the swap file was world-readable. *www3.ca.com/securityadvisor/vulninfo/Vuln.aspx?ID=3607*
- SecurityFocus has a good description of this problem on Mac OS X 10.3. *securityfocus.com/archive/1/367116/2004-06-24/2004-06-30/0*

Part V: On the Wire

16 Spoofing and Man-in-the-Middle Attacks

In This Chapter

- Finding Spoofing and Man-in-the-Middle Attacks
- Preventing Spoofing and Man-in-the-Middle Attacks
- References

Secure communication is about trust. Unlike access control and privilege, which are intended to allow only an authorized user to gain access to protected information, communication implies that two parties are involved. Ideally, those parties have a mechanism in place to ensure that the information can be passed only to the known, intended recipient. Such a mechanism could take several forms: a shared secret (such as an encryption key), knowledge of the characteristics of the other party (such as a hardware address), or a dedicated communication channel (physical or virtual such as IPSec). Not all communications can take this form, however. The Internet and networks like it work the way they do in part because the participants do not need to know about each other in order to communicate—a publicly available Web page can be downloaded from any machine with a Web browser; likewise, someone with knowledge of another person's e-mail address can send him mail without first exchanging any kind of credential. This "open" nature of communication ensures maximum participation is possible on the Internet; however, such a model of communication has inherent problems. In addition, in the early days of the Internet, security of communications was not a significant concern because the networks were not open to wide, public use.

This chapter deals with two security problems that arise from the way information is handled and exchanged via the Internet: spoofing and man-in-the-middle attacks. These vulnerabilities allow an attacker other than the intended recipient to gain access to information during communication. In *spoofing*, the attacker disguises (or "spoofs") his identity or other characteristics to match that of the intended recipient. In a *man-in-the-middle attack*, an attacker uses his unique location on the network, either between the two communicating hosts or sharing a common network path with one of them, to surreptitiously obtain information.

FINDING SPOOFING AND MAN-IN-THE-MIDDLE ATTACKS

Spoofing is possible in any communications mechanism that relies on the identity of the remote communicator to be secure but doesn't provide a means for positively asserting that communicator's identity. It is also possible whenever that identity is not periodically checked or rechecked with each successive communication. The most common kind of spoofing people think of is IP spoofing, in which an attacker makes his machine appear as though he has an IP address other than the one assigned to him. This is useful to him in several circumstances. First, in communications that do not require a reply from the recipient to reach the sender, any IP address can be used. All the attacker needs to do is to create an IP packet with the correct checksum by hand and write it out to his Ethernet device with the MAC address of the router as the destination MAC. The router routes the packet to the real destination, but the packet shows the spoofed source if it is inspected at any point along the way.

Though spoofing through a router has very few uses (especially now that most of the low-level networking bugs have been found in major operating systems), the same technique can be used for a variety of purposes on a local network. If the attacker shares a LAN where sniffing is possible, he can engage in full two-way communication with the victim host. This is because the victim attempts to reply to the spoofed address by sending a response through the router. Because the attacker is on the same LAN, he can see the response and reply again as though he were the spoofed system. He needs to do this faster than the router can determine that the "spoofed" system does not exist and broadcast an "ICMP Unreachable" message to the victim; however, because of the relatively fast speed of local LANs compared to routed networks and the Internet, this is usually not a problem. Figure 16.1 illustrates this kind of spoofing.

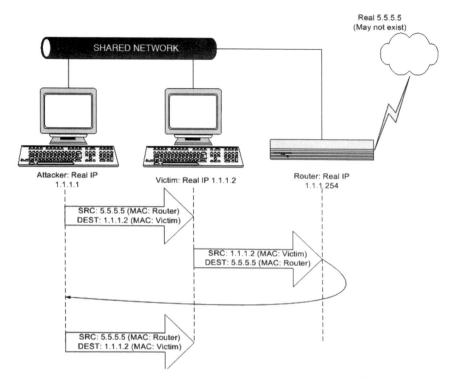

FIGURE 16.1 Local IP spoofing.

This kind of spoofing is successful at evading one kind of security limitation imposed on communications—host-based inclusion and exclusion. Inetd, the networking daemon in Unix, uses two files, hosts.allow and hosts.deny, to limit connections to network services to specific hosts. A system administrator who wanted to allow connections only from a specific host could "deny all" and allow only that specific host to be connected. Many network services such as FTP, Samba, and NFS implement similar host-based restriction lists. Worse still, protocols such as RLOGIN and RSH allow shell access and arbitrary command execution to be granted to a remote user based on his host identity. This means that an attacker can execute arbitrary commands on another host configured in such a manner, as long as he is on the same network as that host.

Using a technique known as ARP cache poisoning, an attacker can accomplish a man-in-the-middle attack using IP spoofing. Like the previous example, this attack relies on the fact that an attacker is on the same LAN as the victim. This attack is possible because the Address Resolution Protocol (ARP), used to associate IP addresses with Ethernet addresses, does not authoritatively check the identity of a host before storing a cache entry. Even though IP-to-Ethernet associations rarely change, ARP does not take this into consideration, so an attacker can send spoofed ARP packets to make himself the middleman in communication between a local machine and the router. Here's how it's done:

1. First, an attacker sends a spoofed ARP packet with his own MAC address to the router, effectively telling the router that he is the owner of the victim's IP address. The router updates its cache to reflect this, meaning incoming packets destined for the victim's IP address are transmitted over Ethernet frames destined for the attacker.
2. Second, the attacker sends a similar spoofed ARP packet to the victim with his own MAC address and the router's IP address, fooling the victim into thinking that he is the router. Outgoing packets destined for the router's IP address from the victim's machine are sent to the attacker's machine instead.
3. When the victim attempts to transmit something to a machine that is not on the local network, all of its traffic is redirected through the attacker's machine. As long as the attacker periodically responds to ARP requests in the correct manner, and correctly routes all traffic from the router to the victim and vice versa, this is transparent to the victim.

The result is that any plain text data transmitted to the Internet from the victim's machine is visible to the attacker, regardless of whether a switched network is in place. Worse, the attacker can use this technique to perform session hijacking attacks and to spoof the identities of other machines on the network from the victim's point of view. For example, the attacker could make it appear as though a Telnet server actually running on his own machine is instead a server that the victim logs into to check e-mail. When the victim types his password into the fake server, the attacker now has that password. This technique works with a variety of protocols that use unencrypted transmission, including some of the major database server and file sharing protocols. Figure 16.2 illustrates ARP cache poisoning as a means of accomplishing a man-in-the-middle attack.

Connection Hijacking

Connection hijacking is when an attacker gains control of a communication midstream and uses that control to either disrupt or resume the communication. The simplest way to gain control of a connection is, you guessed it, spoofing. Simple TCP connection hijacking can be used as a denial-of-service attack on local net-

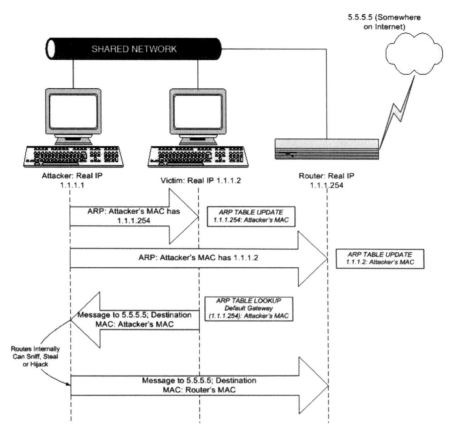

FIGURE 16.2 ARP cache poisoning.

works and networks where a man-in-the-middle attack is possible. To reset a connection, all an attacker needs to do is send a packet with the reset (RST) flag set with the correct sequence and acknowledge numbers and the spoofed source IP of one of the communicating parties. The other party immediately closes the connection in response.

TCP connection hijacking works because, even if the data between two endpoints is encrypted, the TCP headers are not. Thus, the source and destination IPs as well as sequence and acknowledge numbers are always visible to the attacker if he can see the traffic. This same technique can be used to gain control of a connection without closing it. As long as the attacker can prevent the victim from communicating any more responses, he can effectively trick the remote computer into thinking he is the victim's computer. Two methods are used to do this. In the past, when vulnerabilities such as teardrop allowed an attacker to crash the victim's TCP stack, effectively silencing the machine on the network, an attacker would use an exploit such as this to cease communication from the victim. He was then free to resume the connection by spoofing the victim's IP. Because not very many machines are vulnerable to this kind of attack anymore, another method, SYN flooding, can also be employed. The victim's TCP stack can accommodate only a limited number of SYN connections (connections in a half-open state) before it has to stop accepting connections to process replies. Because the attacker is on the same local network as the victim, he can typically send many more of these half-open connections than a victim can respond to. This process is called SYN flooding. By SYN flooding the victim, he can make the victim stop responding to the remote system long enough to resume control of the communication. Figure 16.3 illustrates connection hijacking.

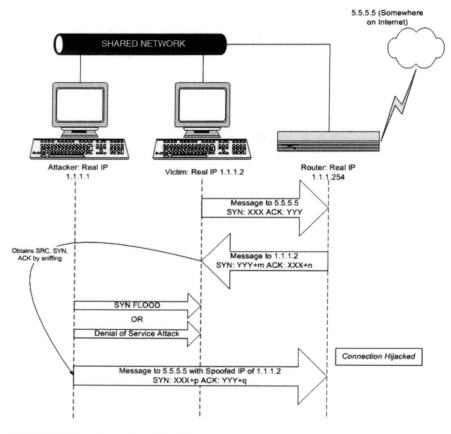

FIGURE 16.3 Connection hijacking.

Connection hijacking can be used for a number of purposes. By hijacking a Telnet connection, an attacker gets a shell that is already logged in as the victim user. This shell can be used to change the user's password, install root kits and back doors, or for any other malicious purpose. By hijacking an HTTP connection, an attacker can insert malicious JavaScript into the HTML stream from a remote server. He can also modify executable files downloaded via HTTP, FTP, or a file-sharing protocol to contain viruses or Trojans. By hijacking a database server connection, an attacker might be able to run queries against the system for information he otherwise wouldn't have access to. Needless to say, any of these techniques can be used to steal information, modify or delete files, and cause miscellaneous havoc.

Hijacking an SSL Session with Ettercap

Ettercap, by Alberto Ornaghi and Marco Valleri, is a utility that can perform a wide variety of cache poisoning, session hijacking, and man-in-the-middle attacks. Ettercap is available for download from *http://ettercap.sourceforge.net/*. The Ettercap user interface was rewritten for version 0.7 in July 2004; however, we use version 0.6.b in our example.

In its simplest form, Ettercap works by ARP cache poisoning the victim and the victim's gateway. By making the victim think that the machine Ettercap is running on is the router, and by making the router think that the Ettercap machine is the victim, traffic between the two (essentially all Internet-bound traffic from the victim) can be sniffed or modified by Ettercap. Ettercap works best on Linux, and in our example we installed it on a Linux machine on the same network as our victim machine. Ettercap can work in either a hub or switch environment.

You might have considerable difficulty if you try to reproduce this example by running Ettercap on the victim machine or on the gateway.

When Ettercap is launched, via the `ettercap` command, it scans the local subnet for active IP addresses. It identifies all of the machines within the local LAN, provided they are on the same subnet as the Ettercap machine. The first step in SSL hijacking is to select a source and destination IP address. Ettercap performs ARP cache poisoning on both systems in an attempt to "connect" them through itself.

The Ettercap user interface presents two columns of IP addresses. The left column is used to select the source machine, and the right column the destination machine. Using the arrow keys to navigate, we select the victim machine (the machine whose secrets we want to steal) as the source machine. This is because the victim is initiating the connection to the remote, SSL-protected server. When we press "Enter," the source selection appears in the top window. Figure 16.4 shows source selection within the Ettercap user interface.

244 The Software Vulnerability Guide

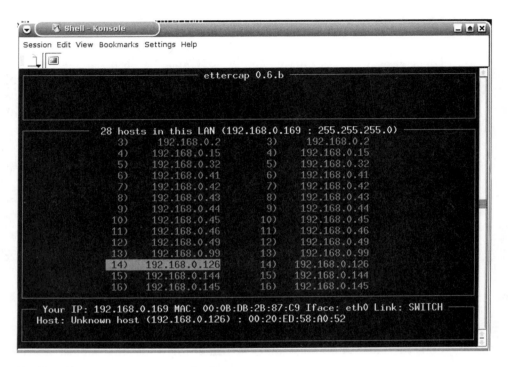

FIGURE 16.4 Ettercap source selection.

Once the source is selected, we select the IP address of the border router as the destination IP address. This is typically an IP address ending in either 1 or 254, but can be another address. To find out, you can use the `netstat` command in Windows or Linux:

```
C:\>netstat -rn

Route Table
===========================================================================
====
Interface List
0x1 ........................... MS TCP Loopback interface
0x2 ...00 50 56 c0 00 08 ...... VMware Virtual Ethernet Adapter
(Network Address
 Translation (NAT) for VMnet8)
0x3 ...00 50 56 c0 00 01 ...... VMware Virtual Ethernet Adapter (basic
host-only
 support for VMnet1)
```

```
0x4 ...00 20 ed 58 a0 52 ...... Realtek RTL8139 Family PCI Fast
Ethernet NIC -
Packet Scheduler Miniport
===================================================================
===================================================================
Active Routes:
Network Destination        Netmask          Gateway       Interface
Metric
          0.0.0.0          0.0.0.0      192.168.0.1   192.168.0.126
20
        127.0.0.0        255.0.0.0        127.0.0.1       127.0.0.1
1
      192.168.0.0    255.255.255.0    192.168.0.126   192.168.0.126
20
    192.168.0.126  255.255.255.255      127.0.0.1       127.0.0.1
20
    192.168.0.255  255.255.255.255    192.168.0.126   192.168.0.126
20
Default Gateway:        192.168.0.1
===================================================================
Persistent Routes:
  None.
```

The default router is the IP indicated as `Default Gateway`, which is also the gateway for destination `0.0.0.0`.

We select the destination IP address from the second column in the same manner we selected the source address. The destination IP address also appears in the top window in Ettercap. Figure 16.5 shows this.

Once the source and destination are selected, we can ask Ettercap to perform the ARP cache poisoning. To do this, simply type the letter "A." Ettercap injects and, after the poisoning is complete, begins sniffing traffic on the network. In the meantime, we proceed to open an SSL-protected Web site in the browser on the victim's machine. Immediately we notice a difference: the browser provides us with an SSL man-in-the-middle warning. This is because the browser is capable of detecting when an improper certificate is used to establish trust between the client and server. A well-educated victim might notice this and choose not to proceed. However, even advanced users are sometimes "conditioned" to ignore these messages, as

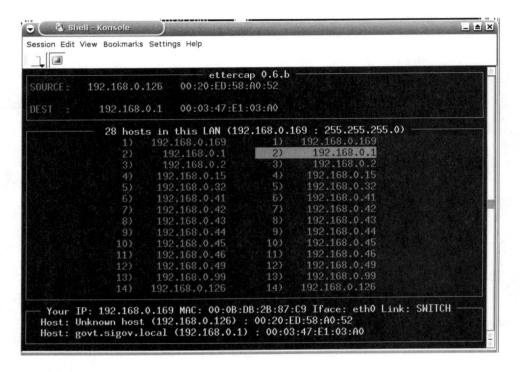

FIGURE 16.5 Ettercap destination selection.

SSL certificates often expire or are configured incorrectly. Additionally, some systems use valid, unsigned certificates to avoid paying the registration fee for a signed certificate. Figure 16.6 shows the results of our browsing attempt in Firefox.

Assuming the user proceeds, the login page for our SSL-protected mail server loads, as is shown in Figure 16.7. The user can type in his username and password as usual. However, Ettercap is now brokering the communication between the victim and the remote server. It can also record everything the victim is transmitting.

To view the contents of the SSL-protected session, we scroll in the Ettercap sniffer view until we find the connection. In our example, we attempted to connect to the remote host, `mail.sisecure.com`, whose IP address is 69.44.157.137. Figure 16.8 shows the Ettercap sniffer interface, with the session directed toward port 443 (HTTPS) of this system.

By pressing Enter, Ettercap shows the recorded contents of the SSL session. Notice that the text of the session (in this case, our mailbox) is plainly visible in the right-hand window, as shown in Figure 16.9.

Spoofing and Man-in-the-Middle Attacks   **247**

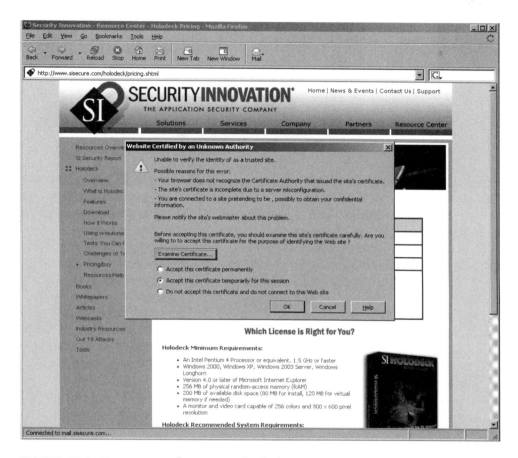

FIGURE 16.6   Browser spoofing attempt in Firefox.

Name Server Cache Poisoning

The attacks we've looked at so far involve spoofing at the IP layer. IP spoofing, as we've seen, usually requires adjacency to the victim on the network to work. It also requires that the attack be carried out on a specific victim, with knowledge of the victim's IP and sometimes MAC address. *Name server spoofing*, on the other hand, can be used to trick whole domains of systems that are dependent on the same name server for address resolution. Remember, most services and users access remote systems by domain name, not IP; this name is resolved to an IP address before the sender initiates communication. By spoofing the domain name of a remote system, we can make the victim think the attacker's IP is the IP of the system he is trying to communicate with.

248 The Software Vulnerability Guide

FIGURE 16.7 SSL-protected mail login screen.

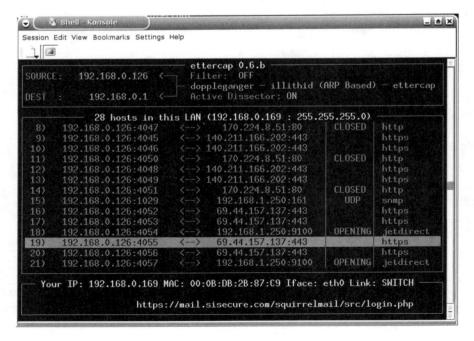

FIGURE 16.8 Ettercap sniffer interface.

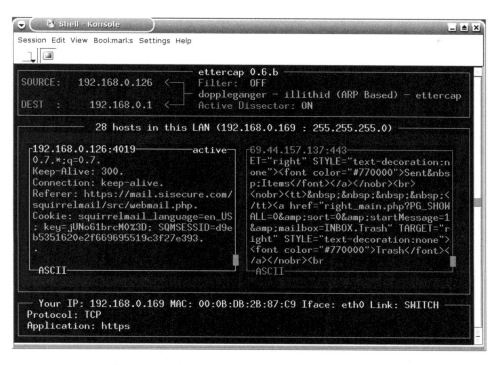

FIGURE 16.9 Mailbox text visible in Ettercap.

DNS spoofing relies on the victim performing lookups from a name server that is vulnerable. A vulnerable server is one that allows updates from its root name server without verifying the identity of that server. Anybody can push DNS updates to a vulnerable server. Because not every server is vulnerable (many ISPs require their downstream name servers to use secure DNS), this vulnerability is more likely to be used by an attacker looking for a target of opportunity such as a spammer or a person trying to carry out a "phishing" attack.

SPOOFING DOMAIN NAMES AND "PHISHING"

"*Phishing*" (pronounced identically to "fishing") is a social engineering attack in which the attacker impersonates a bank, credit card company, or other legitimate authority to extract passwords or financial information from a victim. "Phishermen" cast out a net in the form of a mass e-mail appearing to be from

→

the bank, asking the users to e-mail their account information, go to a Web site and re-enter it, or other such scam. Those who fall for the scam become targets of opportunity for the scammer.

Spoofing of domain names lends tremendous legitimacy to phishing scams. If an attacker can make his fake login page appear to be in a subdomain of the bank he is impersonating, for example, the user has no reason to believe (apart from common sense) that the e-mail's claims are illegitimate. Likewise, having a spoofed e-mail server to respond to inquiries allows the scammer to comfort a "mark" that might initially raise questions.

Because DNS cache poisoning can be used to spoof IPs to multiple users, it increases the likelihood that the scam attempt will succeed in at least one instance. Because the cost of perpetrating one of these scams is very low, an attacker does not have to have many successes to prevail.

Presently, bugs in the Internet Explorer Web browser that incorrectly display partial URLs, or overlay those URLs with other text, are the main mechanism phishermen use to spoof the domain name to their victims. However, as these bugs begin to be fixed, we expect to see DNS cache poisoning become a more popular means of perpetrating these scams.

Spoofing at the Application Level

Vulnerability to spoofing is not confined to the IP layer only. Application-level protocols are subject to spoofing as well. Perhaps the most famous example of this is Simple Mail Transport Protocol (SMTP), which provides no mechanism for verifying that the sender address of an e-mail is legitimate. (Modern SMTP clients can verify the host from which the connection originated, but this can be spoofed by any of the techniques in this chapter and provides no protection against one user in the same domain spoofing another. For example, anyone at *sisecure.com* can send e-mail as anyone else at *sisecure.com*.) This vulnerability has led to the greatest security menace of the Internet age: spam. If you're like the average e-mail user, you've probably received an unsolicited e-mail message since you started reading this chapter. Spammers spoof the sender address of the mail to disguise its real origin—if the real domain name a spam originated from were known, it would quickly be blocked. Spammers use an open relay, a mail server that does not validate hostnames, to accomplish this spoofing.

Spoofing the sender on an open relay is simple—just supply a bogus e-mail address. It's that easy. Really. Figure 16.10 illustrates communication with an open relay e-mail server.

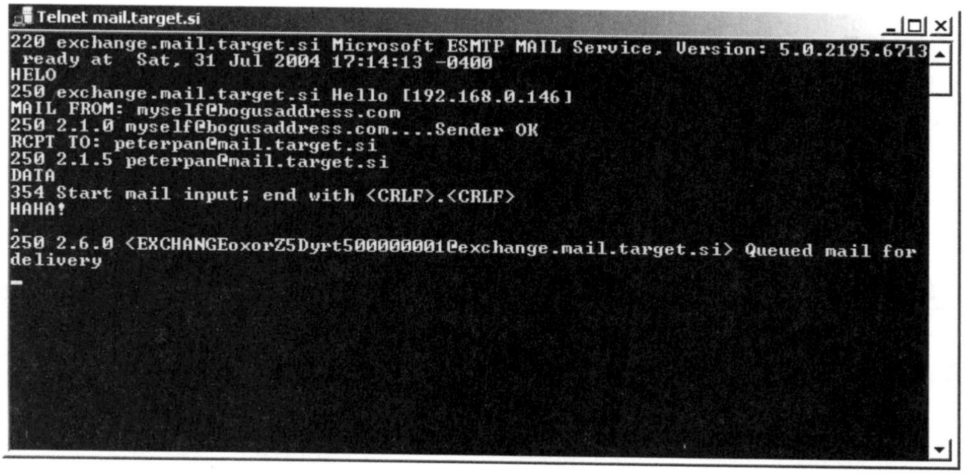

```
Telnet mail.target.si
220 exchange.mail.target.si Microsoft ESMTP MAIL Service, Version: 5.0.2195.6713
 ready at  Sat, 31 Jul 2004 17:14:13 -0400
HELO
250 exchange.mail.target.si Hello [192.168.0.146]
MAIL FROM: myself@bogusaddress.com
250 2.1.0 myself@bogusaddress.com....Sender OK
RCPT TO: peterpan@mail.target.si
250 2.1.5 peterpan@mail.target.si
DATA
354 Start mail input; end with <CRLF>.<CRLF>
HAHA!
.
250 2.6.0 <EXCHANGEoxorZ5Dyrt500000001@exchange.mail.target.si> Queued mail for
delivery
```

FIGURE 16.10 Spoofing the sender (SMTP).

The Common Gateway Interface (CGI) mechanism used for Web application programming supports a number of environment variables including the HTTP_REFERRER variable, which is used to indicate the previous URL that "referred" the user to the requested page. While this variable was intended to be used primarily to track which external links to a page were used to reach it, Web designers have been using the variable to prevent forceful browsing and other state-based attacks on Web pages. This is dangerous, because CGI relies on information supplied by the client browser to set the HTTP_REFERRER variable. An attacker with access to a modified browser (easy to do via the MSHTML object in Windows) or a Web proxy can change this variable to any value he wants. As a result, CGIs should not depend on the referrer to determine whether a request has been securely and legitimately made.

Other Kinds of Man-in-the-Middle Attacks: DHCP and 802.11

It is considerably easier to accomplish man-in-the-middle attacks in an environment where IPs are dynamically assigned. For example, the Dynamic Host Configuration Protocol (DHCP) is used to dynamically assign IP addresses to machines within a local area network. By spoofing MAC addresses, the attacker can obtain all of the available IP addresses for a LAN, effectively blocking additional machines from being able to get on the network. Additionally, he could set up a rogue DHCP server that sets the default gateway of its clients to his own IP, permitting man-in-the-middle attacks.

802.11 networks are even more susceptible to man-in-the-middle attacks. This is because on 802.11 access points, all systems are within the same collision domain—that is to say, an 802.11 network looks like one big hub. Even with Wired Equivalent Privacy (WEP) this is possible because it is necessary only that we be able to broadcast data to an arbitrary node on the collision domain to perform ARP cache poisoning; it is not necessary that we be able to sniff the traffic. Access points can even propagate these vulnerabilities to the wired network to which the system is attached. This is especially serious because it solves the "adjacency" problem. A machine no longer needs to be plugged into the same switch as its victim; it can be located anywhere the wireless network is accessible.

PREVENTING SPOOFING AND MAN-IN-THE-MIDDLE ATTACKS

The key to preventing spoofing and man-in-the-middle attacks is to successfully authenticate the remote system. Secure Sockets Layer (SSL) does this by taking information we can know about a transaction, such as the IP address and hostname of the remote system, and encrypting it with a private key. These keys are known only to the certificate authorities that issue SSL certificates—VeriSign is the largest of these in the United States. The public key associated with the certificate is published and is known by every browser that provides SSL support. A Web site owner is issued a certificate only for his own hostname and IP address; it is not portable to other systems. If the certificate information does not match the actual information for the host, a warning message is produced in the browser.

As we have said previously, HTTP is not the only protocol on the Internet, and SSL is not the solution to every problem. Many other applications, such as SSH, the Secure Shell protocol, are implemented similarly, though. Just as strong cryptography is the only solution for protecting data from being read, certificates are the only foolproof solution for preventing man-in-the-middle attacks. Implementing a certificate system is relatively challenging, and we do not recommend it to the reader. An interested person should check out some of the better cryptography books, such as Bruce Schneier's *Applied Cryptography* [Schneier95], before undertaking such a task.

Summary Sheet—Spoofing and Man-in-the-Middle Attacks

Problem:

TCP/IP does not provide built-in capability to positively identify a remote system. Because a packet passes through many networks from source to destination, an attacker in the middle of this route, or adjacent to one of the endpoints, might be able to exploit this problem to steal information or to hijack an in-process connection.

→

Potential Impact:

An attacker could, using these attacks, view a communication that he is not authorized to view. In addition, he might be able to "hijack" or take control of one side of a communication and impersonate one of the communicating parties.

Habitat:

This problem occurs frequently in software where data must be protected in transit. Examples include programs that exchange usernames and passwords, encryption keys, or personally identifiable information, such as a credit card number, over an open network.

Tools You Need to Find It:

One of the best tools out there for session hijacking and man-in-the-middle attacks is Ettercap (*http://ettercap.sourceforge.net/*). It is capable of sniffing on switched and unswitched networks; hijacking common plain text protocols such as telnet, SMTP, POP, and NNTP; and intercepting SSL-encrypted Web traffic.

How to Look for It:

Identify all of the protocols associated with an application. Are they plain text? Do they exchange authentication information? If so, do they do so in the clear? How is the remote system positively identified?

Symptoms of Failure:

The protocol does not use encryption or challenge/response to determine the authenticity of the remote system. The protocol does not have any facility to report whether a man-in-the-middle attack has taken place.

Famous Failures/Exploits:

- *www.securityfocus.com/bid/3460* describes a man-in-the-middle vulnerability in the 802.11 wireless protocol.
- *www.thoughtcrime.org/ie-ssl-chain.txt* describes a vulnerability in Internet Explorer that could allow an attacker to perform a man-in-the-middle attack without the user's knowledge.

REFERENCES

[Schneier95] Schneier, Bruce. *Applied Cryptography: Protocols, Algorithms, and Source Code in C*, 2nd Ed. Wiley, 1995.

17
Volunteering Too Much Information

In This Chapter
- Finding This Vulnerability
- Fixing This Vulnerability

Whenever an application must respond to an error condition, the developer has several critical choices to make, such as what information about the error message should be revealed to the user and what information about the application or system configuration and state it should contain. On its face, this appears to be a pretty straightforward decision for the developer: if the cause of the error is known to the error handler, then include it in the error message. If the error was caught by some global error handler and the specific cause is unknown, return a general error message. These guidelines make sense from a development point of view because the more diagnostic information returned to diagnose the problem the better. This holds true for "planned" failure scenarios such as the failure to connect to a remote server, failure to authenticate, and so on. Problems arise, however, when error messages reveal information about a system or application that can be leveraged by an attacker. The problem here is that often an application reveals too much information about itself, its state, user data, its host system, or the network. In many cases, an attacker can gain a significant advantage by interpreting and using the information gained from application and system error messages.

As an example, consider the default configuration of many Web servers with an insecure Web application running. One of the most common attacks against Web applications that interact with a database is SQL injection. With SQL injection, an attacker attempts to enter SQL commands into an input field with the hopes of changing the database query that is constructed at the server end. If the server-side code is written so that the query is constructed directly from user data (a common practice—see Chapter 21 for more details) and has no error handling code, an attacker might be able to modify the SQL query. In most cases, an attacker must make assumptions about what the server-side query looks like, and thus launching a successful attack can involve a fair amount of trial and error while refining the input string. When the SQL statement is syntactically incorrect, the default behavior in many cases is for the Web server to return an error message that contains important details about the server. Figure 17.1 shows one such error message.

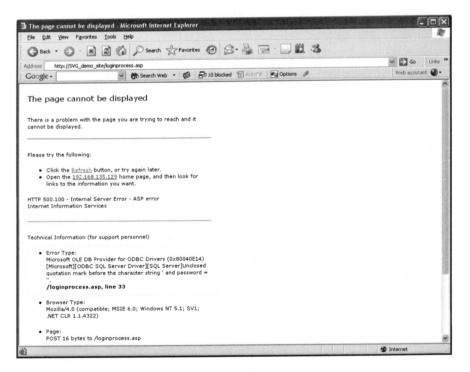

FIGURE 17.1 Web site showing ODBC error.

In the highlighted portion of Figure 17.1 we see that several important details about the server are revealed. First, we now know that Microsoft SQL server is running, which means that an attacker can try to exploit older vulnerabilities on that system that might have been carelessly left unpatched. The error message also reveals some of the server query string and lets us know, for example, that one of the table column names is "password," which can be incredibly valuable for future SQL injection attacks. The most important information revealed, however, is that this Web application is indeed open to SQL injection. Figure 17.2 shows another error message that was generated from a different input string that was provided to the application. This error message is actually telling the attacker how to fix his attack string so that it works.

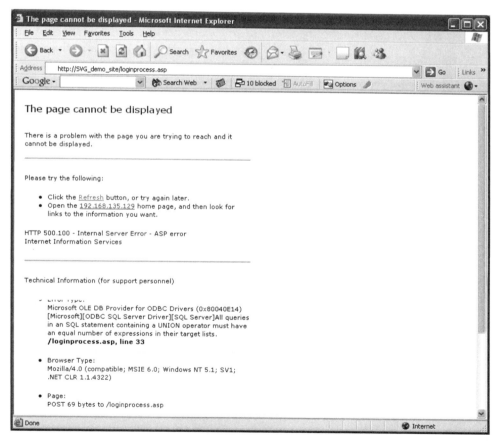

FIGURE 17.2 ODBC error showing results of SQL injection.

Error messages such as these give valuable information to a developer who is trying to debug his code, which would certainly make the behavior appropriate during testing of the application. For production systems, however, this information is of little use to the legitimate user and is grossly insecure in the information it reveals to a would-be attacker. These types of information leakages abound in software. Some of the more interesting instances are where information can be inferred from a series of messages. One interesting example can be found in the Quick 'n Easy FTP Server 1.77 published by Pablo Software Solutions (*www.osvdb.org/displayvuln.php?osvdb_id=3574*). The software allows users to connect remotely and upload and download files. A security vulnerability exists, however, in the way it returns error messages using the del (delete) command. Assuming that the user has no delete permissions on the server, when a specified file exists, the application returns the error message "550 Permission Denied." When the file does not exist, the software returns the error message "550 File not Found." Using this disclosure, an attacker can not only determine the existence of a specified file, but also can use it to gain valuable information about the target system, such as the underlying OS and which patches are installed (through the existence of specific fix logs or files).

Another interesting example comes from the popular file sharing utility Samba (available from *www.samba.org*). Samba has a Web management interface called SWAT that requires a user to authenticate to it, as shown in Figure 17.3.

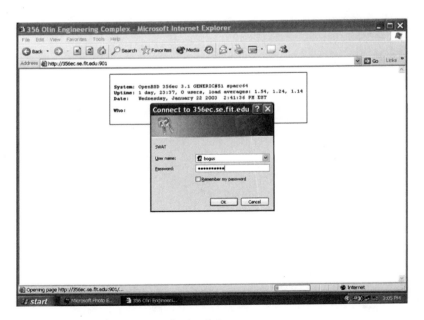

FIGURE 17.3 Samba SWAT login dialog.

When an invalid username is entered, the Web server immediately returns the error message "401 Bad Authorization: username/password must be supplied," as shown in Figure 17.4.

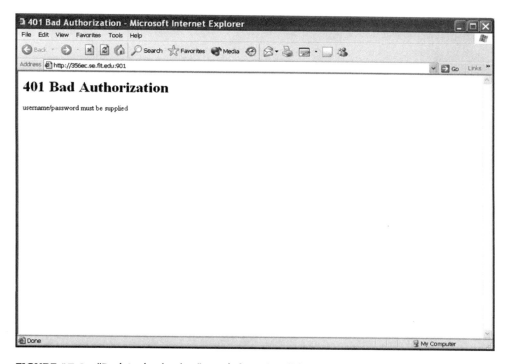

FIGURE 17.4 "Bad Authorization" result from invalid username.

When a valid username is entered with an invalid password, after a short pause, the error message "401 Authorization Required: You must be authenticated to use this service" is returned, as shown in Figure 17.5.

In and of themselves, each error message is fairly good from a security point of view: they are fairly generic and do not give an attacker any additional information about the system. When taken together, however, an attacker can discriminate between a valid and invalid username. Using a dictionary attack, an attacker can now enumerate all valid accounts for the application. An error as subtle as this can dramatically decrease the security of the system. An attacker can now concentrate on what he knows to be valid accounts and focus on cracking or brute forcing those accounts.

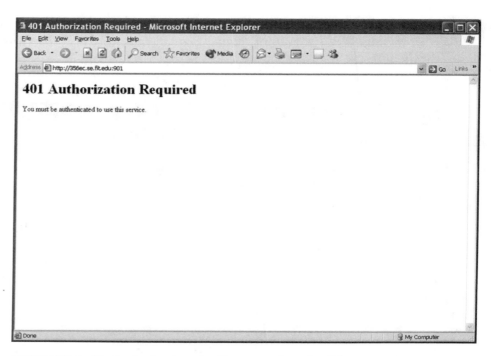

FIGURE 17.5 "Authorization Required" message from invalid password.

Information need not be disclosed solely through error messages, however. Sometimes, information about a system can be inappropriately discovered from handshakes, login prompts, or greetings. A careful balance must be drawn between enhancing the user experience and creating a system weakness.

FINDING THIS VULNERABILITY

We can broadly decompose an application into two kinds of code: functional code that provides the core functionality of a system and error handling code that keeps the functional code from failing. When we design a system, typically most of the effort is put into architecting features and how those features and the functions that support them interact. This functional code is usually exercised fairly extensively through testing and many development paradigms focus on testing for the presence of correct behavior. Now consider error handling code. Error handlers are subjected to far less exposure to testing than the functional code they are designed to protect. Error handlers are also frequently written in response to errors that occur during use and are thus not given the same level of design scrutiny as functional code. Error handlers might, therefore, be written at different times to handle similar situations as errors occur. All of this adds up to code that is written with a very narrow focus to handle very specific situations, and it is easy to understand mistakes such as two slightly different code paths yielding different error messages and thus disclosing information about the software's operation.

From a code review standpoint, enumerating application strings and scanning them individually for parameters that might contain sensitive information can be a good starting point for uncovering individual revealing error messages. Another good code-based technique is to look for generic error message strings and then find out which code paths reference them. The rule of thumb is that there should be a minimal set of such error messages and all relevant error code should point to the same strings as opposed to repeating those strings in multiple locations.

From a black-box testing perspective, it is critical to force applications to exercise their error handling code and traverse paths that expose messages to the user. Errors that are returned to remote users are the most critical, but local errors must be scrutinized based on where the risks for your particular applications lay. For example, applications that run as a higher privilege level than their users might be able to be manipulated into disclosing information about privileged system files or user files.

FIXING THIS VULNERABILITY

When writing error messages, creating welcome banners, or generally interacting with users the inclination is to be as descriptive as possible. Descriptions that are helpful to a user, such as error messages that identify common user errors or causes, are appropriate and can help someone interact better with the system or diagnose and fix a user entry problem. The rule of thumb, however, is to not disclose information that is not available from other sources and is unnecessarily informative. For example, a typical good error message for failed authentication contains common tips such as the use of CAPS lock. A bad error message might tell the user that the username was correct but the password was wrong or what the system policy is on passwords (e.g., passwords must contain a maximum of eight characters with numbers and letters only). In the latter case, the benefit to an attacker far outweighs the benefit to the user of disclosing this information. When writing error messages, or providing information to a user in general, you should consider the following:

> **Does the user have the minimal information that he needs to take corrective action?** A good error message tells the user some potential remedies to the problem. An authentication error message, for example, might suggest checking to see if CAPS lock or NUM lock is on.

Does this message provide more information about the system than is available from other legitimate sources? A secure error message should not reveal any information that is not legitimately accessible from other sources. Consider, for example, the error message returned by imagemap.exe on an old and unpatched version of Microsoft Windows 98 with personal Web Server. Figure 17.6 shows that the error reveals the path on the server to the Web root. While this is the default path, this information can be incredibly valuable to an attacker.

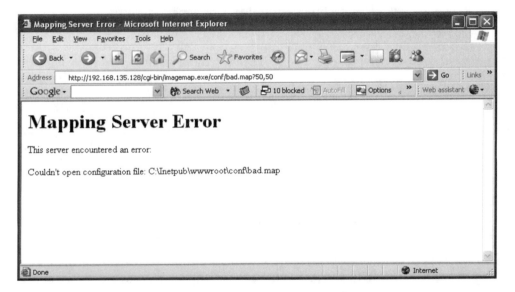

FIGURE 17.6 Error message exposing information about the organization of the filesystem on the server.

Is each piece of information disclosed necessary/helpful to the legitimate user? Consider again the ODBC error message returned by a server in Figure 17.1. Here the user is bombarded with information about the type and version of the database server along with information about the SQL query that resides on the server. This information is completely useless to a legitimate Web user entering information into a Web form, but can be immensely helpful to an attacker.

Does the corrective advice in the error message indirectly disclose sensitive information? Some error messages can reveal sensitive information by the remedies they suggest. One example would be a failed login message that tells a user that passwords are a maximum of eight characters, which are all alphanumeric. This information is mildly helpful to the user, but it is far outweighed by the information it discloses to the would-be attacker.

When giving information to a user, careful consideration must be given to the potential advantage this information could give to an attacker.

Summary Sheet—Revealing Too Much Information

Problem:

Whenever we create an error message or provide information to a user, our initial inclination is to be as descriptive and prescriptive as possible. The problem is that information about the system or other sensitive data might be disclosed that can be useful to an attacker.

Potential Impact:

Information leakage can be a serious security concern, especially when it gives an attacker an inroad into a system. Error messages can be exceptionally helpful to an attacker, especially when they return information about the filesystem, other user data, system configuration, or other data that the viewing user should not have access to.

Habitat:

Any application. The problem is especially severe in applications that are accessible through the network.

Tools You Need to Find It:

A keen eye!

How to Look for It:

Looking for information disclosure in source code entails tracking down error strings and making sure that the information revealed to the user is informative but minimal. From a black-box testing perspective, it is critical to expose error handling code to testing and view each error message at least once to determine if the information disclosed to the user is necessary and appropriate. In addition to single error messages, always check for information disclosed through comparing different error messages.

Symptoms of Failure:

Symptoms can be very subtle. Look for technical details in error messages and banners that reveal information about the system or can give an attacker an advantage in breaking into a system.

Famous Failures/Exploits:

- **CAN-2004-0050:** Verity Ultraseek before 5.2.2 allows remote attackers to obtain the full pathname of the document root via an MS-DOS device name in the Web search option, such as (1) NUL, (2) CON, (3) AUX, (4) COM1, (5) COM2, and others.
- **CAN-2003-0512:** Cisco IOS 12.2 and earlier generates a "% Login invalid" message instead of prompting for a password when an invalid username is provided, which allows remote attackers to identify valid usernames on the system and conduct brute force password guessing, as reported for the Aironet Bridge.

Part VI
Web Sites

18 Cross-Site Scripting

In This Chapter

- Finding Cross-Site Scripting Vulnerabilities
- Fixing This Vulnerability

Our first Web site vulnerability, cross-site scripting, is presented because it is one of the easiest to exploit. While some attacks require sophisticated knowledge of system and assembly programming, or the ability to execute many test cases before finding a vulnerability, cross-site scripting is literally accomplished by cut and paste. It is often the first attack a hacker tries against a Web site to test the general security of the site. Fortunately, it is also one of the easiest attacks to prevent. Despite this, cross-site scripting vulnerabilities continue to be found in a number of sophisticated sites and Web-based applications.

Cross-site scripting is essentially injecting malicious HTML, including Web-based scripting like JavaScript, into the input of a Web-based application, in the hopes that the application might print that code back out again unparsed. When the dynamic page with the malicious code is output by the Web server, the browser attempts to render it as though it were HTML, complete with scripting capability.

Like format string attacks, the vulnerability results from the system allowing *data*, in this case HTML, to be mixed and interpreted with *code* such as JavaScript. What makes cross-site scripting worse than format string attacks is that the browser can execute a broad set of arbitrary commands using JavaScript, so no complicated exploit string is required.

Take a look at an example from "SI Jams and Jellies," a mock e-commerce site built using MySQL and Perl. The site contains a search page that allows patrons to search for jams or jellies that match a certain keyword. Figure 18.1 shows the search page.

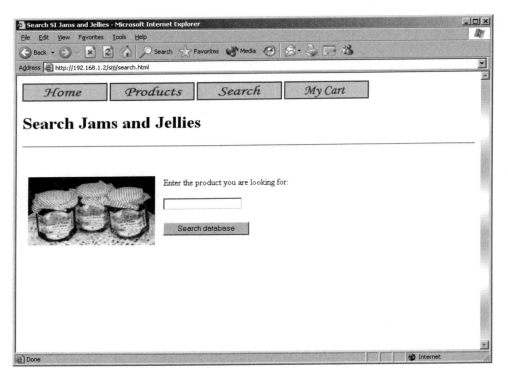

FIGURE 18.1 SI Jams and Jellies search page.

When a patron types a keyword in the form field and clicks "Search Database," the data is passed to the `searchJJ.pl` CGI program. This program queries the database and returns any hits associated with the keyword. Figure 18.2 shows the results returned from a search for the keyword "mint," which is rendered in a table showing product, description, and price.

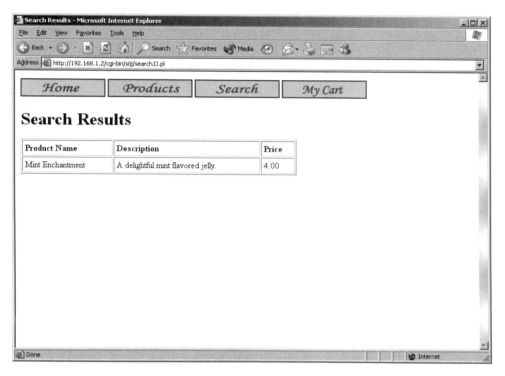

FIGURE 18.2 Results returned from search page.

So far, we've looked at the result of a successful query. What happens when we search for a keyword that isn't likely to be found? Figure 18.3 shows the results of a query for the keyword "dinosaur."

No table is returned; the CGI simply responds, "No information for dinosaur." However, this result is more interesting than the previous one from the point of view of cross-site scripting. Nowhere in the previous query (apart from in the rows of the result set) did the keyword we search for appear in the dynamic page returned by the CGI. However, searching with other erroneous keywords quickly confirms that the CGI *always* prints the keyword upon an unsuccessful search. What are the chances that this CGI is vulnerable to cross-site scripting? To be vulnerable, the CGI would have to respond blindly with whatever search term we supplied it, even if that term contained HTML or JavaScript. If we can substitute "dinosaur" with a malicious script, the system is vulnerable.

Most frequently, cross-site scripting vulnerabilities are used by an attacker to obtain information, such as a cookie or session ID, which is available via the Document Object Model. However, some cross-site scripting vulnerabilities might

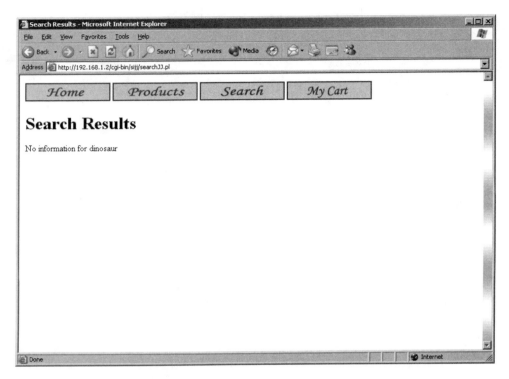

FIGURE 18.3 Results for search of keyword "dinosaur."

allow execution of arbitrary code on the victim's machine, especially when the vulnerability is in a trusted site or Web-driven client application, or is combined with an unsafe ActiveX control.

Cross-site scripting can also be used to:

- Deface a Web site. By inserting script that modifies the content of a page, or pops up an alert, an attacker can vandalize another person's Web site.
- Manipulate the Document Object Model. If scripts can be executed, any of the functions available through the browser's Document Object Model can be called, including functions that read and write files (usually blocked in "unsafe" domains, but cross-site scripting bugs are not confined to these), pop up windows, or manipulate cookies and history.
- "Poison" cookies by modifying them to suit an attacker's purpose. Chapter 20 describes cookie poisoning in more detail.

FINDING CROSS-SITE SCRIPTING VULNERABILITIES

Let's take a look at a piece of JavaScript code. The `<script>` tag is used to embed JavaScript (or another script language) within an HTML page. When the browser encounters this tag during rendering of a page, it executes the JavaScript commands contained within the tag. JavaScript works with a set of objects called the *Document Object Model*. These objects can be used to manipulate elements of the browser, document, and page within which the script is contained. For example, the `location` attribute of the `window` object (`window.location` in JavaScript) is used to manipulate the URL (Web site location) of the current page. Setting `window.location` to a different address causes the browser to navigate that window to a new page. One useful feature of JavaScript is the `alert()` function, which causes a message box (Windows `MessageBox` object) to appear with the text specified in the argument. So the script:

```
<script>
 alert(window.location);
 </script>
```

would pop up a message box with the address of the current page loaded in the browser.

Let's assume that our search CGI is not vulnerable to cross-site scripting. In all likelihood, we see a message, "No information for <script>alert(window.location);</script>," or perhaps an error message. However, if it *is* vulnerable, when we type the preceding script into the search field, it pop up a message box with the address of the CGI page, "*http://192.168.1.2/cgi-bin/sijj/searchJJ.pl.*" Figure 18.4 shows the results of our malicious test query. As you can see, the message box with the URL information is popped up immediately after rendering "No information for," even prior to loading the rest of the page. Clicking on the OK button within the message box causes the rest of the page to load, as illustrated in Figure 18.4.

The effects of this cross-site scripting attack are relatively benign. In our example, a user can inject the script only into a page that is returned to him in his own browser. The real danger is that a malicious user could leave a cross-site script behind for another user to stumble on; this would be the case if the script were posted to a message board, Web log, or collaboration site. A cross-site script in the "User Reviews" section of an online bookstore or in a posting on an auction Web site would likely reach a broad audience. Because JavaScript can do a great deal besides pop up a message box, an imaginative attacker could create scripts that carry out a variety of malicious actions. For example, a vulnerability existed in an online auction site. (We do not suggest that any particular site is vulnerable; the folks at the major auction companies are well aware of this problem.) It's possible that the

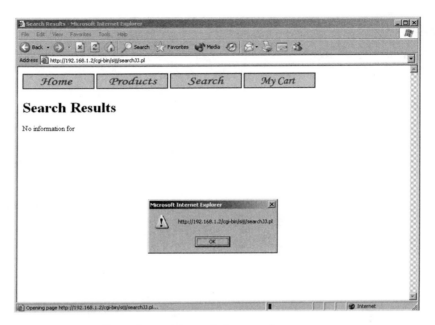

FIGURE 18.4 Effect of cross-site scripting attack.

user's identity is verified from page to page by means of a session cookie stored in the user's local cookie store. JavaScript is allowed to retrieve the value of cookies set by pages in the same domain in which the script is running. An attacker could post a cross-site script that retrieved this session cookie and posted it as a CGI parameter to a third-party (malicious) Web site. When an ordinary user accesses the malicious auction page, *his* session cookie is stolen and posted to the hacker's site. Armed with this cookie, the attacker can impersonate the user, making bids, purchases, or changing personal information, for as long as the session cookie is valid. If the auction is a particularly popular one (as auctions for Beanie Babies were in the early days of eBay and Yahoo! Auctions), the attacker is continuously updated with a fresh supply of session cookies, so he need only try the most recent ones in hopes of finding a cookie that hasn't expired.

The problem of cross-site scripting is even more serious in pages in older browsers, that don't have restrictions on execution of malicious ActiveX controls like `Scripting.FileSystemObject`, and in pages that a user is automatically redirected to. A vulnerability existed in Versions 4, 5, and 5.1 of Microsoft Internet Information Server (IIS), one of the most popular Web server applications on the Internet. A malicious user could craft a special URL that, when requested from any IIS server, would inject a cross-site script into the 404 page returned by the server.

The "Famous Failures/Exploits" listing mentioned in the Summary Sheet section of this chapter contains a pointer to this vulnerability. This means that a malicious site could redirect a user to a page that executed arbitrary script as though it were in the domain of the remote server. This could be used to steal cookies belonging to that domain, or to execute ActiveX controls if that domain is a member of trusted sites.

CIRCUMVENTING CLIENT-SIDE INPUT VALIDATION

Some developers think the solution to preventing malicious input is validation within the JavaScript of the page itself. Don't do this. It's like having students grade their own tests. The reason is that it's relatively easy to undo client-side input validation. A developer can validate input in an HTML form field in only a few ways. They are:

MAXLENGTH: This sets the maximum number of characters that can be typed into input fields. This is sometimes the only protection provided against buffer overflows within Web applications.

onblur: This event fires when a user moves the mouse out of an input field or presses the Tab key. This is the most common input validation routine.

onchange: This event fires when a user types anything within the input field. It can be used to exclude certain undesirable characters.

onkeydown, onkeyup, onkeypress: These events occur when the user types a key within the input field.

onfocusout: Similar to onblur, this fires after a user has performed an action that changes focus (such as hitting the Tab key), but before focus is lost.

Any of these techniques can be disabled automatically by a skillful attacker. The MAXLENGTH field cannot be disabled (easily) but can be set to any arbitrary length. A MAXLENGTH of 65,536 would likely permit buffer overflows to be found in the server application, if there are any. The onblur event and other events are relatively simple to disable. Here is some Visual Basic code that disables all onblur events in a page:

```
For x = 0 To WebBrowser1.Document.Forms.length - 1
    For y = 0 To WebBrowser1.Document.Forms(x).elements.length - 1
        WebBrowser1.Document.Forms(x).elements(y).onBlur = ""
    Next
Next
```

FIXING THIS VULNERABILITY

The root cause of a cross-site scripting vulnerability is usually in the CGI component of the Web application. Let's take a look at the portion of Perl code that processes our search request:

```
$dbh = DBI-> connect(
  "dbi:mysql:database=si;host=localhost;port=3306",
  "root",
  "sisecure")
  or die("Couldn't connect");
$query->import_names('R');
$search=$R::searchJJ;
$sth = $dbh->prepare("SELECT * FROM products where" .
  "products.pname like '$search%' or products.pname like" .
  "'%$search'")
or die("Error");
$sth->execute
    or die("cannot execute");

if($sth->rows == 0)
{
    print "No information for " . $R::searchJJ;
}
else # display the table of results
```

Based on our observations with the browser, the problem is in the line that prints the "No information for X" message. The variable $R::searchJJ is obtained directly from the name-value array passed into the program from the search page's form field; Perl performs no parsing on this apart from translating it from the URL-encoded string transmitted on the wire back to the original source data. If $sth->rows == 0, we print $R::searchJJ explicitly. When the browser attempts to render our output, it interprets the contents of $R::searchJJ exactly as they were supplied.

This problem can be avoided by parsing $R::searchJJ before printing it back out. If our only concern is "breaking" cross-site scripting, Perl's regular expression capability can easily help in this. The easiest way would be to change the greater than and less than symbols to their HTML literal entity equivalents:

```
$search =~ s/</&lt;/g;
$search =~ s/>/&gt;/g;
```

Preventing More Advanced Cross-Site Scripting Vulnerabilities

Often times an attacker "encodes" a cross-site scripting vulnerability using an alternate representation of the attack string. URL encoding, C, and Perl escape strings and Unicode encoding are popular techniques for bypassing cross-site scripting filters. Depending on the language and stages of filtration, any of the following could represent the `<script>` tag in HTML:

```
%3cscript%3e            (URL-encoded)
%u003cscript%u003e      (Unicode %u-encoded)
%%3cscript%%3e          (Double URL-encoded)
\074cscript\076         (C/Perl Octal Escape Characters)
\x3cscript\x3e          (C/Perl Hex Escape Characters)
```

The problem is even worse in applications that accept UTF-8 characters, because you have multiple ways to encode each character. Quickly, the number of combinations that would need to be filtered becomes unmanageable. In this case, it is better to delete characters that do not explicitly conform to a safe, allowable set. The command:

```
$search =~ tr[_a-zA-Z0-9 ,./!?()@+*-][]dc;
```

can be used to do this.

HTML-Encoding Output

Microsoft's Active Server Pages has a function, `Server.HTMLEncode()`, that can be used to encode output for safe display in HTML. `HTMLEncode` translates an arbitrary string into the HTML representation of that string. Essentially, it translates the special characters used to form tags, scripts, etc., in HTML into their literal printable equivalents. By doing this, you can be assured that script code that made it through as input cannot subsequently be output without first being converted to "safe" formatting. If you don't use ASP, it's relatively easy to write your own encoder. The characters you need to translate are:

- & should become &
- < should become <
- > should become >
- " should become "
- (should become (
-) should become)
- # should become #

Summary Sheet—Cross-Site Scripting

Problem:

This problem arises from the reuse of input from HTML form fields as output without parsing. A malicious user could insert script code into an HTML form field, which gets interpreted by the browser when that data reappears as output in a Web page.

Potential Impact:

An attacker could, in certain circumstances, execute arbitrary JavaScript code on the client computer. This technique is most frequently used to steal cookies that contain personally identifiable information or are used as tokens to gain access to a Web site.

Habitat:

CGI programs running on a Web server are susceptible to this vulnerability. All platforms and languages are vulnerable in one way or another, because all allow arbitrary script code to be written back to the browser. Any user who permits JavaScript to be executed in Web pages is vulnerable to this attack.

Tools You Need to Find It:

A Web browser and a sample JavaScript program, such as the `alert(window.location)` script, are needed to find simple cross-site scripting bugs. More sophisticated forms of the attack might require the ability to encode Unicode, UTF-8, or C-style escape strings.

How to Look for It:

Insert the sample JavaScript program into HTML form fields (including hidden ones) and observe the results.

Symptoms of Failure:

The program executes the script behavior at the point during page loading where the output text is rendered.

Famous Failures/Exploits:

- Numerous examples of cross-site scripting vulnerabilities in Web sites, applications, and operating systems can be found on *Securityfocus.com*.
- *SecuriTeam.com* has a description of the Internet Explorer universal cross-site scripting vulnerability at *www.securiteam.com/windowsntfocus/5QP0A206VK.html* and the IIS vulnerability at *www.securiteam.com/windowsntfocus/5WP0J006UG.html*.

19

Forceful Browsing

In This Chapter

- Description
- Finding Forceful Browsing Vulnerabilities
- Preventing Forceful Browsing

Web browsers, all things considered, are relatively simple programs. In their most basic application, they listen on a specified port (port 80 for regular HTTP; port 443 for HTTPS), receive requests for files, and transmit the contents of those files back to the requestor on the same port. Where security is needed, it comes in one of several forms. First, the Web server is typically prevented from copying files for which it does not have appropriate permission. In a multi-user system, files owned by a user other than the one that launched the Web server are blocked unless the owner gives the Web server user permission to read them. The Web server also accesses only pages reachable from its document root, typically the htdocs directory in Apache and inetpub/wwwroot directory in IIS. So, taken from the Web server's point of view, the files it is able to serve are relatively safe: an administrator must give permission for the Web server to read a file, and even then it reads only files in directories that have been configured to be used with the Web server. That is to say, unless a CGI program comes into play. And because most all Web applications have the CGI interface at their core, it's almost guaranteed to.

This vulnerability affects both traditional CGI programs (written in Perl or C) and applications written with an application server framework such as J2EE, ColdFusion, or ASP.NET.

DESCRIPTION

As we said before, the Web server accesses only files reachable from its document root directory. However, CGI is intended to allow ordinary programs to interact on the Web browser interface, and ordinary programs can access any part of the filesystem, not just the document root. File permissions might prevent the access of some important system files, such as the /etc/shadow file in Unix. However, many important system files, such as the ordinary passwd file, are configured to be read by an arbitrary user, including the Web server user. These files would not be automatically off limits to a CGI program. To make matters worse, the Web server can usually run only as one system user. This user is nobody on most Apache systems, and IUSR on IIS. This is true regardless of the identity of the remote user requesting the page. As a result, all human users, whether they are administrators, ordinary users, or attackers, are grouped together as nobody for purposes of file permissions. It is possible, via a set of HTTP configuration files separate from OS file-level permissions, to assign permission to a directory to a specific set of Web users. However, remember these permissions are not applied to files accessed via a CGI program. The Web server can restrict *access* to a CGI to specific users, but cannot limit the files that CGI can access.

This fact gives rise to a number of vulnerabilities, called *forceful browsing* vulnerabilities. They are called this because an attacker can, by manipulating values in the data submitted to the CGI program via the browser, force the Web server to return Web pages or other files that the attacker would not otherwise have permission to access in the application's security model. Consider a simple example. In our example Web site, SI Jams and Jellies, the administrator is allowed to update the pricing in the product catalog by means of an admin page. The link to this page is made visible only to the administrator when he logs in. Figure 19.1 shows the admin page.

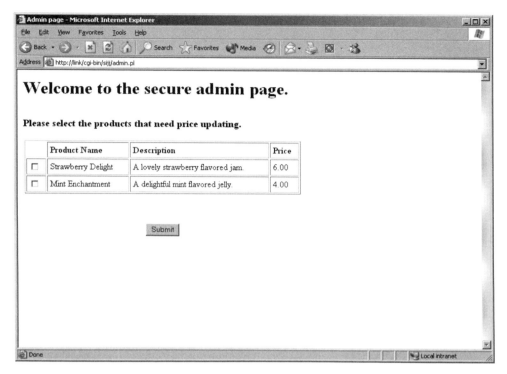

FIGURE 19.1 SI Jams and Jellies admin page.

What prevents an attacker from accessing this page to update the pricing for a product in a way that benefits him? (A negative price might actually allow a credit card to be charged back when the item is purchased.) Because a link to the admin page is not displayed anywhere in the site visible to a normal user, he has to guess the URL. But that isn't too difficult considering what the page is named. If the attacker guesses the name and browses directly to /admin.pl, he has a good chance that he might be able to update the prices.

This scenario represents the simplest form of forceful browsing. Unfortunately, it is one of the most commonly found vulnerabilities in Web sites. In this case, the attacker would have been limited in the damage he could cause to the functionality exposed by /admin.pl, and only if he guessed the name of the admin page. However, this same technique can be used to gain inappropriate access in situations when the URL is known. Consider a scenario in which a user buys a trial subscription to a Web site. If he saves the "deep" URLs made visible during the subscription period, he might be able to cancel his subscription and still access the content. Or imagine an e-commerce site that produces a temporary, intermediary URL during the processing of a credit card refund. Will replaying this URL cause the card to be refunded a second time?

If the attacker is able to cause the card to be refunded a second time, he has accomplished a *session replay attack*. Session replay attacks are a kind of forceful browsing that exploits a fundamental problem in Web applications: state. Because HTTP was designed primarily to serve arbitrarily requested files, the concept of state was not very important to protocol designers. Even CGIs were not meant to be accessed in a stateful fashion. The problem came when people wanted to imitate the functionality of conventional interactive programs that can store stateful information between inputs from the user. Because it doesn't know the identity of the next user who might access the CGI program, it can't tailor the functionality based on that user. Thus the *user* must supply the CGI all of the information needed to respond to the request. To pass stateful information from CGI to CGI, hidden fields and cookies are frequently used. However, both hidden fields and cookies are subject to tampering by the user.

In this way, the attacker can trick the CGI program into thinking that a transaction that has already happened hasn't happened yet. He can also make the CGI think he's someone he's not. Often, Web application programmers use a variable called the session variable to associate an authenticated user with a particular session within the Web site. Because the Web is not real time, these session variables cannot immediately expire. If an attacker is able to obtain the session variable of another user, he can impersonate that user without knowing his username or password, because the session variable makes the Web application think the user is already logged in. Session variables are often easily obtained by attackers; they are either stored in hidden fields within the HTML pages associated with the site, or are transmitted on the URL. While SSL can be used to prevent attackers from accessing another user's session variable, SSL isn't always used on all portions of a site where the session variable is visible. It also might not be useful in circumstances where the session data is predictable. Additionally, cross-site scripting attacks (Chapter 18) can be used to obtain the session variable from the user's cookie cache or browser history.

Forceful browsing needn't be confined to URLs. URLs are certainly the easiest places to inject forceful browsing information, as the injection mechanism, the URL bar in the Web browser, is accessible to any user. However, forceful browsing can also be accomplished through tampering with CGI parameters found in hidden fields (or even unhidden ones). Suppose we had a CGI that dynamically included a text file to present content to a user, as is sometimes done in sites with a large amount of content. For example, a dictionary CGI might read the contents of "definitions/dinosaur" to display a definition when the user requests one for the word "dinosaur." The actual URL might look like this:

    http://www.dictionary.site/definition.cgi?word=dinosaur

and the CGI might be a C program that parsed the parameters and called

```
char *env, *word, filename[1024];
FILE *fv;

env = getenv("QUERY_STRING");
word = strchr(env, "=") + 1;
snprintf(filename, 1024, "definitions/%s", word);
file = fopen(filename, "r");
```

to open the definition file. An attacker could pass the parameter

    word=../../../../../../../../../etc/passwd

and use this CGI to read the contents of the `passwd` file, or any other file for that matter.

Consider this vulnerability that existed in ColdFusion Server 2.0 through 4.0 from Macromedia. ColdFusion is a popular Web application server platform. (For more information about ColdFusion, you can visit *www.macromedia.com*.) A utility CGI, `exprcalc.cfm`, which shipped with ColdFusion software, allowed a Web developer to test ColdFusion expressions. However, this utility was accessible by an ordinary user in many default installations of the product. The script relied on another CGI, `openfile.cfm`, to access a file that had been uploaded for evaluation. The filename of this file was passed to `exprcalc.cfm`, which evaluated the file and then deleted it. However, `exprcalc.cfm` never checked that the source of the filename passed to it was `openfile.cfm`. As a result, an attacker could forcefully browse to `exprcalc.cfm` to force it to delete an arbitrary file. To make matters worse, the attacker could use `exprcalc.cfm` to delete `exprcalc.cfm`, meaning the uploaded file was left permanently on the server.

FINDING FORCEFUL BROWSING VULNERABILITIES

Fortunately, finding forceful browsing vulnerabilities is a much easier for the developer than the attacker. The developer does not need to guess the names of admin pages and can roam around the filesystem as he pleases. The technique for finding these vulnerabilities depends on the kind of forceful browsing we're looking for.

To find hidden admin pages, first create a list of "safe" URLs that should be accessible to the whole world, including users who haven't logged in. Also, ask yourself the questions, "how difficult is it for an attacker to get a minimally functional user account? Does the application create a new account for anyone who completes a form? Does any information supplied by the applicant actually get verified?" It is typically much easier for an attacker to accomplish forceful browsing if he has a minimal set of credentials for a system.

Log into the site as an administrator and manually walk through all of the pages, noting each URL. A Web spider can also be used to do this automatically. After this is done, log out of the site completely, and clear the browser cache to ensure no pages are still cached. Do not log in as a regular user just yet. Attempt to re-access each of the saved URLs. Are any accessible? If so, are they on the list of safe URLs? Remember, these URLs can be accessed by *anyone*, not just an authenticated user. Now repeat the process, this time logging in as an ordinary user. Again, try to re-access the saved URLs. Are any of the administrator pages accessible?

Testing for session replay attacks is done in relatively the same way. Identify how your application stores session information. Is it done via a hidden field in each page? A parameter passed on the URL? Log in as an ordinary user and navigate to a "deep" URL—one that is not accessible without logging in. Note the URL in the navigation bar; does it contain any CGI parameters with names like "session," or very long strings of random numbers and letters? If so, this is likely the session variable.

Save the current page to a file and log out. Open the saved file in the browser and click on one of the links. Can you still navigate around the site? If so, you are vulnerable to a session replay attack. What happens if you paste the URL into the navigation bar? Will that allow you to access the site without logging in? Finally, test your session replay attack from another machine. If you can still access the site without logging in, it's likely that an attacker can as well.

Identifying forceful browsing vulnerabilities in parameters is a bit more difficult. In our dictionary example, the attacker had no special knowledge that the definitions were stored in files. They could have as easily been stored in a database, or in a one big file that is parsed at runtime. But an attacker focuses his attention on parameters that are obviously the names of files—parameters that contain directory browsing characters like "/" and "\", parameters that contain obvious file extensions like .gif and .exe, and parameters associated with files a user can upload, download, view, etc. Each place where this occurs in your application, test some common path traversal strings such as "../" and "c:\" and see what the effect is. Be sure to test the URL-encoded representations of these strings as well: "../" can also be "%2e%2e%2f."

> **GOOGLING FOR HIDDEN FILES**
>
> An interesting way to find hidden admin pages, password files and other forceful browsing targets is to actually *search* for them. Many sites support internal searching, either through their own search CGI or through an external index like Google® (*www.google.com*). Often, search CGIs are provided by the application server software or are cut and paste from a Web master's resource site, and do not take security into consideration. As a result, sometimes the admin pages are indexed along with the ordinary content—to the delight of the attacker. Try searching for these keywords to see if they turn up any hidden admin pages or interesting results:
>
> - admin.asp
> - config.asp
> - password.asp
> - admin.aspx
> - admin.pl
> - config.pl
> - admin.cfm
> - /etc/passwd
> - passwd
>
> - shadow
> - .htaccess
> - .htpasswd
> - root
> - Administrator
> - Webmin
> - miniserv.pl
> - password
> - username

Building a Forceful Browsing Test Tool

Microsoft's MSHTML object, which can be manipulated through an ActiveX interface, makes it easy to build a forceful browsing test tool. The MSHTML object gives a developer programmatic access to all of the common functions of Internet Explorer. It is possible to request pages and respond to events, manipulate the Document Object Model of pages loaded in the browser, etc. We chose to build our tool in Visual Basic, because of the ease of manipulation of the MSHTML object using Visual Basic's `WebBrowser` control. Our tool is useful in finding both hidden admin pages and session replay attacks.

Creating the User Interface

The user interface to `ForcefulBrowse` is relatively straightforward. It consists of a `WebBrowser` control (named `WebBrowser1`) with basic navigation buttons, an address bar, and a history list. The navigation buttons (`Toolbar1`) and address bar (`txtAddressbar`) allow a user to interact with the browser control in the same way they would with Internet Explorer, by typing in a URL; going home, forward, and back; and refreshing the loaded page. The history list (`listHistory`) records each URL visited. If parameters were supplied via a `GET` method, those parameters are encoded in the URL. (Our example tool doesn't support `POST` parameters.)

Four buttons to the right of the history list control the testing. The "Gather Links" button (`btnGatherLinks`) parses the page currently loaded into the browser and adds all the links within that page to the history list. Because our tool doesn't have a crawler, this allows us to obtain links by manually navigating to a page and then harvesting them out of the Document Object Model for that page. Why doesn't our tool have a crawler? Crawlers are finicky—they must be able to get around various kinds of authentication, follow links in the correct order to make application logic work, and correctly supply parameters to form fields. In practice, they don't work too well for finding security bugs. (Imagine trying to find a negative total attack, in which an attacker supplies a negative price or quantity in an attempt to cause a credit card to be refunded when there are no items in the shopping cart.) Crawlers have another problem: they often navigate away from the site or domain you're testing. You have some serious legal and ethical implications if a security testing tool follows a link to another person's Web site and begins testing their code for vulnerabilities.

The "Clear History" button (`btnClearHistory`) allows us to empty the history list—pretty straightforward. Ideally we would have functionality to add and remove items from the history list, reorder the list, etc. However, this would serve only to clutter our example; the readers are free to add these features to the tool if they want. The "Test!" button (`btnTest`) starts the testing by iterating through the URLs and forceful browsing scenarios. Lastly, the "Stop" button (`btnStop`) stops the test prematurely if this is necessary. Figure 19.2 illustrates the user interface for `ForcefulBrowse`.

In addition to the visible controls, a Timer control (`Timer1`) is used to throttle the rate of requests for pages. Requesting pages too quickly can bog down both the server and the test tool. The timer is set to 1000 milliseconds.

Building the Basic Browser Functionality

The basic browser functions are relatively easy to implement. By overloading the `Load` event for the main form, we can navigate the browser to the user's home page. This ensures that the home page comes up when the application is launched.

```
Private Sub Form_Load()
    ' navigate the browser to the user's home page
    ' (for appearance purposes)
    WebBrowser1.GoHome
End Sub
```

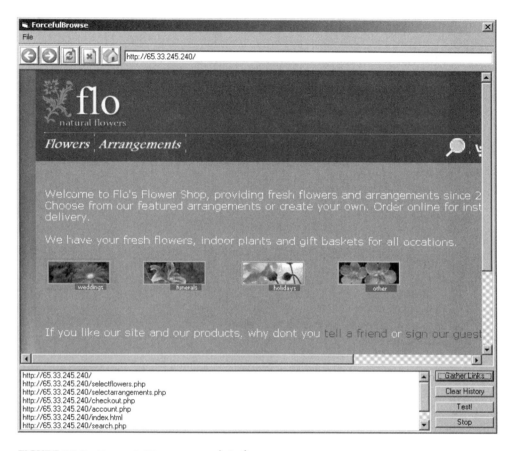

FIGURE 19.2 ForcefulBrowse user interface.

Toolbar1 contains the navigation buttons. A single event handler, Toolbar1_ButtonClick, handles a click event regardless of what button is pressed. As a result, we take a different action based on the button index.

```
Private Sub Toolbar1_ButtonClick(ByVal Button As ComctlLib.Button)

    ' each button within the toolbar has a button index
    ' Button.Index tells us which one was pressed
    Select Case Button.Index

        Case 1:     ' back button was pressed
            WebBrowser1.GoBack
```

```
        Case 2:      ' forward button was pressed
            WebBrowser1.GoForward

        Case 3:      ' refresh button was pressed
            WebBrowser1.Refresh

        Case 4:      ' stop button was pressed
            WebBrowser1.Stop

        Case 5:      ' home button was pressed
            WebBrowser1.GoHome

    End Select

End Sub
```

We want to navigate to a new address when the user presses Enter in the address bar. Because the address bar is a regular text edit control, we can't overload the Change event, because this event is fired on every character typed and has no way of discerning which key was pressed. Instead, we overload OnKeyPress, and navigate if the key pressed was the Enter key.

```
Private Sub txtAddressbar_KeyPress(KeyAscii As Integer)
    ' 13 is the key code for carriage return (enter)
    If KeyAscii = 13 Then
        WebBrowser1.Navigate (txtAddressbar.Text)
    End If
End Sub
```

Finally, we add the code to exit the application if the File>?Exit menu item is selected.

```
Private Sub mnuFileExit_Click()
    Unload Me
End Sub
```

Building the History Information

To capture the URLs navigated to by WebBrowser1, we overload its NavigateComplete2 event. This event occurs when Internet Explorer has successfully (or unsuccessfully) finished trying to navigate to a page opened through the explicit Navigate() method or by clicking on a link.

```
Private Sub WebBrowser1_NavigateComplete2(ByVal pDisp As Object, URL As
Variant)
    txtAddressbar.Text = URL
    If Not testsRunning Then
        listHistory.AddItem (URL)
    End If
End Sub
```

We do not update the history list if the `NavigateComplete2` event happened while tests were running. If we did this, the history list would grow forever as test cases added new URLs to the bottom of the list and then tried to navigate to variations of those URLs using forceful browsing techniques. In addition, we update the address bar's text to contain the URL, because the browser does not do this automatically.

To "Gather Links" from the currently loaded page, we use the links collection within the Document Object Model. This allows us to programmatically obtain all of the links without parsing the HTML associated with the page. While this does not get all links (specifically, dynamic links and those within Java applets and ActiveX controls), any static links within the page are contained in this collection. We iterate across these links in response to the `btnGatherLinks` button press and add them to the history list.

```
Private Sub btnGatherLinks_Click()
    ' Document.links is an array of all the links within
    ' the document
    For x = 0 To WebBrowser1.Document.links.length - 1
        listHistory.AddItem (WebBrowser1.Document.links(x))
    Next
End Sub
```

Clearing the history list is relatively straightforward. We just overload the button's click event and add the code to empty the list control.

```
Private Sub btnClearHistory_Click()
    listHistory.Clear
End Sub
```

Running the Tests

Because our testing is event driven (individual tests are launched and analyzed in response to `Timer` and `DocumentComplete` events), we need some global variables to keep track of where we are in the testing.

```
Dim browserReady As Boolean
Dim testsRunning As Boolean
Dim urlIndex As Integer
Dim testIndex As Integer
Dim resultsString As String
```

The `browserReady` variable is set to false each time we start to navigate to a new page and set to true each time a document is completely loaded within the browser. Because we can only communicate asynchronously with the Internet Explorer control, this prevents the test apparatus from trying to start a new test before the previous one is finished. The `testsRunning` variable is set to true if we're running tests and false if we're not. The code associated with the "Test!" and "Stop" buttons set and unset these, and the test code uses them to determine whether tests are running or not. `urlIndex` is an index into the history list. It tells us what item within the history is currently being tested. Likewise, `testIndex` tells us which forceful browsing technique is being tried. The `resultsString` string stores the results of the testing in report form. This is displayed to the user when testing is completed.

Starting and stopping tests is relatively straightforward. The system launches tests in response to a timer event. To start testing, we need to enable the timer. Once the timer is running, it begins performing tests when the next `Timer` event occurs. Because of this, all we need to do to start testing is clear the results string, reset the indices `urlIndex` and `testIndex`, and set `testsRunning` to true.

```
Private Sub btnTest_Click()
    ClearResults
    Timer1.Enabled = True
    testsRunning = True
    urlIndex = 0
    testIndex = 1
End Sub
```

To stop testing, we set `testsRunning` to false and disable the timer. We also need to set `browserReady` to true, in case we were in the middle of a navigation when the button was clicked. We use `PrintResults` to display the test results in the browser window.

```
Private Sub btnStop_Click()
    testsRunning = False
    Timer1.Enabled = False
    browserReady = True
    PrintResults
End Sub
```

Most of the action takes place in the timer event handler. This handler is called once per second when tests are running. Within this one-second window, we need to do two things: determine whether the last test passed or failed and queue the next test.

Before doing this, we perform some "sanity" checks. If `testsRunning` is false, the user clicked the "Stop" button in between timer events. In this case, we exit, without waiting to analyze the results of the last test. If `browserReady` is false, Internet Explorer failed to completely navigate to the previously supplied URL within our one-second window. If this is the case (and often is when remote servers are involved), all we can do is wait another second and hope for the best.

For every test except the first one, we want to analyze the results of the previous test to determine whether it passed or failed. To do this, we obtain the HTML code to the page from the Document Object Model. The property `Document.body.innerhtml` contains this. We start by assuming that the navigation succeeded. If it did, we don't know what the page contents should look like. On the other hand, if we failed to navigate to the page, we should get an HTML error message back. We compare the page contents to the most common HTML error codes:

- 400 Bad Request
- 401 Unauthorized
- 403 Forbidden
- 404 Not Found
- 500 Internal Server Error
- 501 Not Implemented

If we find one of these codes in a page, navigation likely failed, and we report this in the log. A successful navigation doesn't necessarily imply a vulnerability, but a failed one likely implies that there wasn't one.

The `nextURL` function obtains the next URL to be tested. If this function instead returns the code string, "DONE," we have evaluated the last test case. This is necessary because we need to ensure we evaluated the results of the very last test case before we print the results. If we're "DONE," we call `PrintResults` to display the results string. Otherwise, we queue the next URL to be tested.

```
Private Sub Timer1_Timer()

    ' if the tests aren't running, don't bother to
    ' handle this event
    If Not testsRunning Then Exit Sub

    ' if the browser is not ready, we need more time
    If Not browserReady Then Exit Sub

    ' determine whether the last test passed
```

```
        ' note: we don't need to do this if we're on the
        ' very first test
        If Not (urlIndex = 0 And testIndex = 1) Then
            pageText = WebBrowser1.Document.body.innerhtml

            succeeded = True

            If InStr(pageText, "400") Or InStr(pageText, "Bad Request")
Then succeeded = False ' bad request
            If InStr(pageText, "401") Or InStr(pageText, "Unauthorized")
Then succeeded = False ' unauthorized
            If InStr(pageText, "403") Or InStr(pageText, "Forbidden") Then
succeeded = False ' forbidden
            If InStr(pageText, "404") Or InStr(pageText, "Not Found") Then
succeeded = False ' not found
            If InStr(pageText, "500") Or InStr(pageText, "Internal Server
Error") Then succeeded = False ' internal server error
            If InStr(pageText, "501") Or InStr(pageText, "Not Implemented")
Then succeeded = False ' not implemented

            If Not succeeded Then
                RecordResult ("FAIL: " + WebBrowser1.LocationURL)
            Else
                RecordResult ("SUCCESS: " + WebBrowser1.LocationURL)
            End If

        End If

        ' fetch the next url
        URL = nextURL

        ' if we're done, stop the test
        If URL = "DONE" Then
            testsRunning = False
            PrintResults

        ' otherwise, navigate to the next url
        Else
            browserReady = False
            WebBrowser1.Navigate (URL)
            txtAddressbar.Text = URL
        End If


    End Sub
```

Forceful Browsing

The nextURL function is responsible for composing the URL string to be tested. To do this, it first truncates the working URL (obtained from the history list) to its directory path name. The getURLPath (described in the following) does this. It then appends one of the forceful browsing test strings, based on the test index number. In the event that we've already tried the final test string for this URL, we move on to the next URL. Likewise, if we've tried everything for every URL, we return "DONE" as a code to the test event handler to stop testing and print the results. The test strings are taken from the "Googling for Hidden Files" sidebar in this chapter.

```
Private Function nextURL() As String

    ' if testIndex > number of tests, get the next url
    If testIndex > 16 Then
        testIndex = 1
        urlIndex = urlIndex + 1
    End If

    ' if we found the last url, send a code to
    ' stop the test
    If urlIndex > listHistory.ListCount - 1 Then
        nextURL = "DONE"
        Exit Function
    End If

    baseURL = getURLPath(listHistory.List(urlIndex))

    Select Case testIndex

        Case 1:
        ' try to browse up one level
        nextURL = baseURL + ".."

        Case 2:
        ' try to browse up two levels
        nextURL = baseURL + "../.."

        Case 3:
        ' try to browse up to the root
        nextURL = baseURL + "../../../../../../.."

        Case 4:
        ' try to browse to admin.asp
        nextURL = baseURL + "admin.asp"
```

```
Case 5:
' try to browse to config.asp
nextURL = baseURL + "config.asp"

Case 6:
' try to browse to password.asp
nextURL = baseURL + "password.asp"

Case 7:
' try to browse to admin.aspx
nextURL = baseURL + "admin.aspx"

Case 8:
' try to browse to admin.pl
nextURL = baseURL + "admin.pl"

Case 9:
' try to browse to config.pl
nextURL = baseURL + "config.pl"

Case 10:
' try to browse to admin.cfm
nextURL = baseURL + "admin.cfm"

Case 11:
' try to browse to /etc/passwd
nextURL = baseURL + "../../../../../../../etc/passwd"

Case 12:
' try to browse to .htaccess
nextURL = baseURL + ".htaccess"

Case 13:
' try to browse to .htpasswd
nextURL = baseURL + ".htpasswd"

Case 14:
' try to browse to miniserv.pl
nextURL = baseURL + "miniserv.pl"

Case 15:
' try to browse to admin/
nextURL = baseURL + "admin"
```

```
        Case 16:
        ' try the URL itself (for replay attacks)
        nextURL = listHistory.List(urlIndex)

    End Select

    testIndex = testIndex + 1

End Function
```

The getURLPath function truncates a URL to return only its path, without filename or CGI parameters. To do this, we search backward from the end of the string, looking for the first "/" character.

```
Private Function getURLPath(URL As String) As String
    pos = Len(URL)
    While pos > 0 And Mid$(URL, pos, 1) <> "/"
        pos = pos - 1
    Wend
     getURLPath = Mid$(URL, 1, pos)
End Function
```

The remaining functions within ForcefulBrowse deal with processing the results. We have chosen to implement a very simple results report, which displays "SUCCESS" or "FAILURE" and the URL name for each test case run. This report is loaded into the WebBrowser control upon test completion for the user to review.

```
Private Sub RecordResult(s As String)
    resultsString = resultsString + s + Chr$(13) + Chr$(10)
End Sub

Private Sub ClearResults()
    resultsString = "Test Results" + Chr$(13) + Chr$(10)
    resultsString = resultsString + "————" + Chr$(13) + Chr$(10)
    resultsString = resultsString + Chr$(13) + Chr$(10)
End Sub

Private Sub PrintResults()
    WebBrowser1.Document.body.innertext = resultsString
End Sub
```

Running `ForcefulBrowse`

To find replay attacks and hidden admin pages within your site, follow these steps in `ForcefulBrowse`:

1. Navigate to each page within your application. Allow the history list to record the URLs of each of these pages.
2. Press "GatherLinks" after navigating to a page. This stores the links within the page in the history list.
3. Log out of your application. This should prevent replay attacks from working. If a URL within the site is still accessible after logging out, `ForcefulBrowse` catches this.
4. Click the "Test!" button. You then observe the browser control trying to navigate to each of the pages within the application, as well as the hidden admin pages, password files, etc.
5. Review the results. A "SUCCESS" means `ForcefulBrowse` succeeded in navigating to the page.

Figure 19.3 illustrates the results of `ForcefulBrowse` on our test Web site, *Flowerhacker.com*. Note that it discovered the hidden admin page, `admin/`.

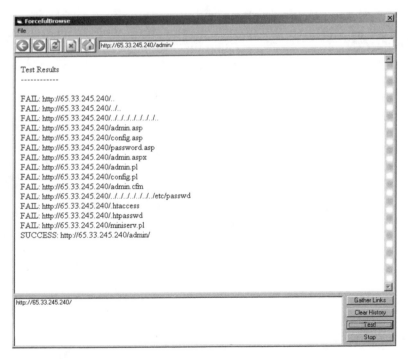

FIGURE 19.3 Using ForcefulBrowse on FlowerHacker.

PREVENTING FORCEFUL BROWSING

Here are some tips for avoiding forceful browsing attacks in your Web applications:

- Check the HTTP_REFERER environment variable for a sane value, and return an error on a referrer you're not expecting. If a page or CGI is supposed to be accessible only by logging into your site, the referrer should be another page within your site. Be sure to cancel the session variable and log the user out, just to be safe. It is possible to spoof the HTTP_REFERER; a sophisticated attacker knows how to do this. However, it is significantly more difficult to do this in an SSL-encrypted session.
- Whenever possible, use the Web server's authentication, not your own. HTTP basic and digest authentication (generally) prevent attackers from accessing a CGI to which you haven't given them permission. In Apache, you can have your application's add user routine dynamically update the .htaccess files when a user is added. Keep all administrator pages in a directory that is not accessible to ordinary users. If possible, use the REMOTE_ADDR variable to limit administrator CGIs to the IP addresses of legitimate administrators, or use a firewall or virtual private network (VPN) to limit access.
- Parse parameters if they are used to access a file. The best way to prevent forceful browsing to another directory is to remove directory characters ("/" and "\") from the parameters. Be sure to parse URL-encoded and Unicode-encoded variations of these characters. If the parameter *must* contain directory names, use chroot() to prevent browsing past the root directory of the Web site.
- Avoid common names for admin pages. These include "admin.php," "admin.aspx," "admin/," "test/," and variations.
- Expire the session variable as quickly as is practical for your application. Consider turning down the default value of half an hour or 24 hours on application server platforms that have a default value. Keep track of the REMOTE_ADDR associated with the session variable. If it changes, expire the session. Cancel session variables when a user logs out.

Summary Sheet–Forceful Browsing

Problem:

Unlike ordinary Web server requests, CGI programs have full access to the filesystem of the machine on which they run. Tampering with the URL and CGI parameters can be used to force the Web server to return a page to which the attacker does not ordinarily have access. Additionally, the filesystem permissions do not necessarily protect a non-administrative user of a Web application from accessing administrative features of the site. It is the application's responsibility to prevent this. Lastly, the use of session variables in hidden fields and URLs could allow an attacker to "replay" that session from another machine.

Potential Impact:

An attacker could access the administrative functions of a Web site as a non-administrative user. Additionally, he might be able to read arbitrary files on a machine, or access functions of a Web site without logging in. In some circumstances, forceful browsing could be used to cause transactions to take place multiple times, or to modify and delete files on the server.

Habitat:

CGI programs running on a Web server are susceptible to this vulnerability. Homemade CGI platforms might be more vulnerable than commercial application servers, though the potential exists for this vulnerability to occur on all servers.

Tools You Need to Find It:

A Web browser is all that is needed to find forceful browsing vulnerabilities, though a Web replay tool and spider make finding them easier.

How to Look for It:

Save the URLs from an administrative session and attempt to replay them. Insert directory browsing commands such as "../" and "\" into URLs and CGI parameters. Replay URLs and saved pages using session variables from a different machine.

Symptoms of Failure:

The program allows a non-administrative user to access an administrative page. The program allows a user to access protected content without logging in. The program can access and return the contents of an arbitrary file.

20
Parameter Tampering, Cookie Poisoning, and Hidden Field Manipulation

In This Chapter

- Cookie Values
- Form Data
- Query Strings
- HTTP Header Tampering
- Finding This Vulnerability
- Fixing This Vulnerability
- References

Web applications work by passing information from a Web browser on a user's machine back to a Web server and vice versa. To return the right content, the Web server is often sent a wide array of information. The first is the GET request in the HTTP header that instructs the server on which page to retrieve. Quite a bit of information is encapsulated in this header, including the page that referred the user to the requested page, the type of Web browser requesting the page, and several other pieces of information. In addition, if that Web application has previously stored values on the client in a structure called a *cookie*, then that information is sent to the server as well. Still, more data is being transmitted. Parameters entered by a user on one Web page—through a Web form, for example—are sent to another Web page by being transmitted to the Web server.

All of this data originates on the client's machine, which means that the user has complete control of these values and can change them arbitrarily before they are sent back to the Web server. For example, consider the Web page shown in Figure 20.1.

298   The Software Vulnerability Guide

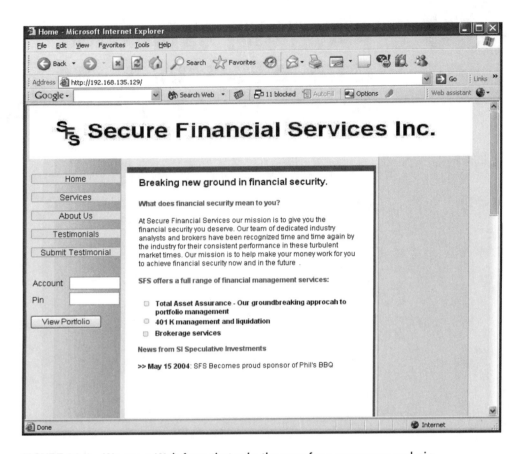

FIGURE 20.1   We see a Web form that asks the user for a username and pin.

In this figure, we see a Web form that asks the user for a username and pin. If we look at the source for the Web page, we can see that our responses are sent to the page "`loginprocess.asp`" using the POST method:

    <form name="form0" method="POST" action="loginprocess.asp">

You have two methods for passing form data to the server: GET and POST. Using the GET method, the form data is actually contained within the URL and is readily exposed to the user. Using POST, that information is sent in the HTTP request to the server after the HTTP header and is not readily visible by the user without the appropriate tools.

If we then enter the username "test" with pin "case", the following information is sent to the Web server:

```
POST /portfolio.asp HTTP/1.0
Accept: image/gif, image/x-xbitmap, image/jpeg, image/pjpeg,
application/vnd.ms-excel, application/vnd.ms-powerpoint,
application/msword, application/x-shockwave-flash, */*
Referer: http://192.168.135.129/index.html
Accept-Language: en-us
Content-Type: application/x-www-form-urlencoded
User-Agent: Mozilla/4.0 (compatible; MSIE 6.0; Windows NT 5.1; .NET CLR
1.0.3705; .NET CLR 1.1.4322)
Host: 192.168.135.129
Cookie: PREF=ID=0d12a6f152d:Locale=815608
Content-Length: 18
Pragma: no-cache
Connection: keep-alive

Acct=test
Pin=case
```

Let's take a minute to look at each line of this:

```
POST /portfolio.asp HTTP/1.0
```

tells the Web server what type or request—in this case, POST—is being made, the page being requested, and the version of the HTTP (HyperText Transfer Protocol) being used. Several types of requests can be made, namely GET, POST, PUT, DELETE, HEAD, TRACE, CONNECT, and OPTIONS, along with non-universally sanctioned extension methods. By far the most common used to request a page are POST and GET. GET does what it sounds like—it requests, or "gets," a specific Web page. POST, on the other hand, is designed to "post" information back to a specific page (from a Web form, for example). POST requests typically contain additional information after the HTTP header, which contains data from the previous page such as values from a form.

```
Accept: image/gif, image/x-xbitmap, image/jpeg, image/pjpeg,
application/vnd.ms-excel, application/vnd.ms-powerpoint,
application/msword, application/x-shockwave-flash, */*
```

tells the Web server which media types are acceptable for response

```
Referer: http://192.168.135.129/index.html
```

tells the Web server which page (URI) this request was made from. This is the page's "referrer" although it is misspelled in the protocol. The `Referer` field is often used to verify that form data was received from the appropriate Web page and not simply constructed by a Web page at a different location containing a modified version of the form. While this is a common check done on the server side, it is trivially bypassable by editing the header information of the request before it is sent. Remember, any data sent from the client can be manipulated.

```
Accept-Language: en-us
```

This limits the response of the Web server to a set of preferred languages—in this case, U.S. English.

```
Content-Type: application/x-www-form-urlencoded
```

This field specifies the type of data being sent to the recipient. In this case, URL encoded form data is being sent.

```
User-Agent: Mozilla/4.0 (compatible; MSIE 6.0; Windows NT 5.1; .NET CLR 1.0.3705; .NET CLR 1.1.4322)
```

The `User-Agent` request-header field contains information about the browser or application making the request. The field can contain multiple product values and also more specific sub-details about the requesting application. In this particular request, the browser has included its type (Mozilla/4.0 compatible), its name and version (MSIE 6.0), the type and version of the operating system (Windows NT 5.1), and the versions of the .NET Common Language Runtime (CLR) installed.

```
Host: 192.168.135.129
```

The `Host` request-header field specifies the server and port number of the resource being requested.

```
Cookie: ID=0d12a6f152d
Jnlnknl
Pragma: no-cache
```

The `Pragma` field has several options that relate to the content. The "no-cache" directive tells the browser to not cache the page's contents and always look to the server for the current version of the page.

```
Connection: keep-alive
```

The Connection field allows the server to set options for a particular connection.

```
Acct=test
Pin=case
```

This is the POST data entered into the form and passed along to the Web server.

In each case, this data was generated by the browser on the client. Any data on the client can be manipulated, and this should not be trusted to make security decisions. Additionally, no data constraints enforced on the client should be assumed to be in place, including header information, cookie values, and user-entered form data. In this chapter we consider the four types of data transmitted to the Web server—cookie values, form data, query strings, and HTTP header information—and the security concerns with each.

COOKIE VALUES

Cookies are small text files that are saved on a user's machine to store information from a Web site. Cookies have become a popular way for a Web application to maintain "state" and keep track of user-specific information. For example, a common cookie use is to store a user ID for a specific Web site. This cookie value is initially set by the site and stored on the user's machine. From then on, every time that user goes to that Web site, the browser checks the machine to see if a cookie has been set for that domain, and if so, it sends the cookie data to the Web server as part of the HTTP header. Several vulnerabilities can result here. The first and most obvious is the storing of sensitive user information within the cookie itself. If another user then has access to the client machine, this cookie data can be stolen. A simple scenario to consider is a Web application that automatically logs users in based on a user ID stored in a cookie. If this ID information is compromised, an attacker can potentially impersonate that user and have access to his account.

A related issue is the tampering of data stored in cookies. Consider, again, the example of a Web site that logs a user in based on an ID stored in the cookie. An attacker could easily iterate through several combinations of possible user IDs and potentially gain unauthorized access to other user accounts. Another issue we ran into recently involved storing the price of items for an e-commerce site in cookies. This particular vendor would allow users to browse the site for items and then add selected items to a virtual shopping cart. The application kept track of what items were stored in the cart by storing their item IDs and corresponding prices on shoppers' machines in cookies (a phenomenally bad idea). The attack vector here is simple: set an arbitrary price for any item you want.

The issue here is trust. Data stored on the client should never be used to make critical security or business logic decisions because of its inherent susceptibility to manipulation.

> **PROTECTING YOUR CUSTOMER'S INFORMATION: YOUR RESPONSIBILITY**
>
> In 2003, California passed an identity theft prevention law that, among other things, requires companies that do business with California citizens to disclose security breaches that might have resulted in loss of a customer's confidential information. Vendors that fail take reasonable precautions to protect confidential information, including credit card numbers and other personally identifiable information, from access by unauthorized persons can be subject to fines, lawsuits, and loss of business.
>
> In April of 2004, the State of New York fined New York City-based *BarnesandNoble.com* $60,000 for exposing customer data through a flaw in their Web site. Personal information (but not credit card numbers) could be accessed, and purchases could be made against another person's account. As best we can discern from descriptions of the bug (including an article on *InternetNews.com* [Naraine04]), this flaw was a simple parameter tampering vulnerability.

FORM DATA

When a Web page accepts data from a user, it can go about filtering it in several ways. A common approach is to use JavaScript running on the client machine to validate data length, content, and type. For example, one could implement such controls to attempt to prevent an attacker from launching a SQL injection attack through the browser by writing the following JavaScript code:

```
checkval=new RegExp("[\-\'\;]");
function validate(){
    if (checkval.test(form0.Acct.value)){
```

Parameter Tampering, Cookie Poisoning, and Hidden Field Manipulation

```
    alert("Account names and passwords should only contain
numbers and letters");
        event.returnValue=false;
    }
    if (checkval.test(form0.Pin.value)){
        alert("Account names and passwords should only contain
          numbers and letters");
        event.returnValue=false;
    }
}
```

and calling that function when the form is submitted:

```
<form name="form0" method="POST" onsubmit=validate();
action="loginprocess.asp">
```

Thus, whenever we enter the characters ";", """, or "-", an error message is raised as shown in Figure 20.2.

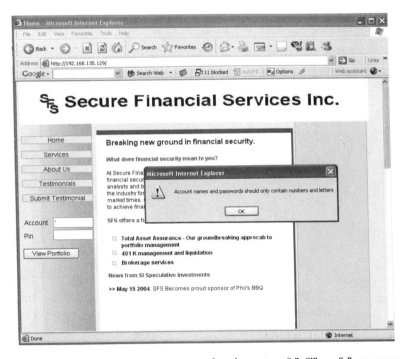

FIGURE 20.2 Whenever we enter the characters ";", """, or "-", an error message is raised.

The issue here is that this check is done on the client side and can easily be bypassed by an attacker. When ensuring the security of data being posted back to the server, it is critical that validation be done on the server. Another issue is data length. A form field can be constrained in HTML, but again this is a constraint enforced by the client that can easily be removed.

Another potentially thorny issue with forms is the use of hidden fields. Hidden fields are exactly what they sound like, form fields that hold values that are completely hidden from view of the user (unless we view the Web page source). Hidden fields in and of themselves are not dangerous, but they are often used by Web developers to transmit values that are assumed to be unalterable by users.

Consider the Web page shown in Figure 20.3. Here we see a "Testimonials" page where users can share their experiences about the company. The testimonials can later be reviewed by the company for possible posting on the Web site. If we take a look at the source for this Web page, however, we can see the hidden field "approved" with value "no" (Figure 20.4).

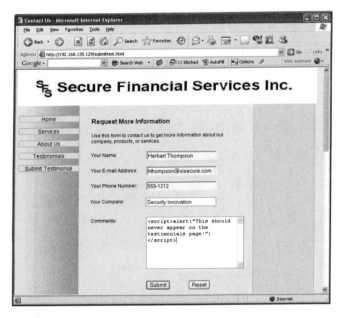

FIGURE 20.3 "Testimonials" page where users can share their experiences about the company.

The data in the "comments" field is JavaScript and should obviously not be approved for posting. One could easily assume, however, that the data entered here is stored in a database and that an approval flag exists that indicates whether or not a particular testimonial has been reviewed and approved for publication on the Web

Parameter Tampering, Cookie Poisoning, and Hidden Field Manipulation 305

```
<td width="180" height="33" colspan="6"></td>
    </tr>
    <tr>
        <td width="324" height="65" colspan="8"></td>
        <td width="28" height="65" colspan="2"></td>
        <td height="117" colspan="2" rowspan="3" valign="top"></td>
        <td width="8" height="65" colspan="2"></td>
    </tr>
    <tr><input name="approved" type="hidden" value="no">
        <td width="508" height="29" colspan="17"></td>
        <td width="8" height="29" colspan="2"></td>
    </tr>
    <tr>
        <td width="324" height="23" colspan="8"></td>
        <td height="29" rowspan="2" valign="top"><input type="submit" value="Submit"></td>
        <td width="16" height="23"></td>
        <td height="29" rowspan="2" valign="top"><input type="reset" value="Reset"></td>
        <td width="88" height="23" colspan="6"></td>
        <td width="8" height="23" colspan="2"></td>
    </tr>
    <tr>
        <td width="324" height="6" colspan="8"></td>
        <td width="16" height="6"></td>
        <td width="240" height="6" colspan="10"></td>
    </tr>
    <tr>
        <td width="750" height="985" colspan="21"></td>
    </tr>
</table>
</form>
</body>
</html>
```

FIGURE 20.4 The hidden field "approved" with value "no".

site. By saving the Web page and changing this value to "yes," we can circumvent this process and "approve" the testimonial, which then gets automatically posted to the site, as shown in Figure 20.5.

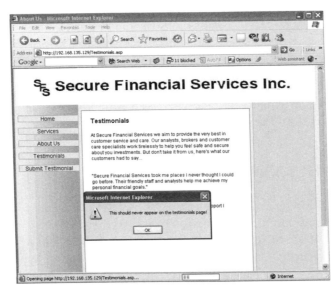

FIGURE 20.5 We can circumvent the process and "approve" the testimonial, which then gets automatically posted to the site.

Public vulnerability databases such as BugTraq and CERT are rife with hidden field manipulation vulnerabilities on commercial Web sites. Don't let your site end up there.

QUERY STRINGS

We've all seen those long URLs that have the name of a Web page followed by a series of "?", "&", and "%" characters. The following is an example from Google when we did a search for "software vulnerability guide":

```
http://www.google.com/search?hl=en&q=software+vulnerability+guide
```

The portion in boldface is referred to as the *query string* and contains information sent from the Web page back to the server. On the server end, the data in this string can easily be parsed and used to return the appropriate response. This URL can be hard-coding on a Web page or constructed by some client-side script dynamically based on data a user has entered into a form. When using a form, all data entered is automatically placed into a query string if the GET method is specified in the form tag.

The query string is interesting from a security standpoint because again the data here is completely at the mercy of the user. Consider the more complicated query string shown here:

```
http://192.168.1.1/a.asp?option=C&V=4&ID=4987201&priv=u
```

A query string with a variety of options is ripe for an attacker to tamper with. A common technique is to tamper with ID numbers and attempt to access another user's account. For some sites we've found that an amazing number of trust assumptions are made on data sent via the query string. Remember, anything that originates from the user should not be trusted.

HTTP HEADER TAMPERING

When a browser sends a request to a server for a Web page, it sends along with it a significant amount of information from the client. Earlier in this chapter we looked at some of the constituent parts of the HTTP header, but a few can have a big impact on security. One of the most interesting is the Referer field. This field tells the server which Web page is making the request for content. For example, if a user clicks on a link on the page *www.herbertthompson.com/index.htm* that points to the page *www.scottchase.com/index.htm,* the Referer tag would be *www.herbertthomp-*

son.com/index.htm. If a user is required to go through a series of pages—filling out a multi-part form for example—a common technique is to look at the `Referer` field and check to make sure that the user is coming from the correct page. The issue here is that the HTTP header information originates from the client and can easily be manipulated. Thus, a Web page that is saved on the local machine and tampered with—to remove some JavaScript validation, for example—can be made to look like it's on the server at its original location.

FINDING THIS VULNERABILITY

Finding these types of issues requires a thoughtful inspection of your Web site. Every time data is passed back to the server, ask yourself what assumptions are being made about that data and how are those assumptions enforced in code. The only filtering mechanisms that can be trusted are those that are securely implemented on the server. A tool called FuzzBrowse can help test for these issues quickly and easily. FuzzBrowse is a Web browser that allows you to look at a Web page and its source code and dynamically change the source code as you are browsing. We developed FuzzBrowse for use internally at Security Innovation®, but have made a version available at *www.scottchase.net/FuzzBrowse/*.

Consider Figure 20.6, which shows FuzzBrowse looking at an online flower shop. This page allows users to pick from a variety of items and then specify and

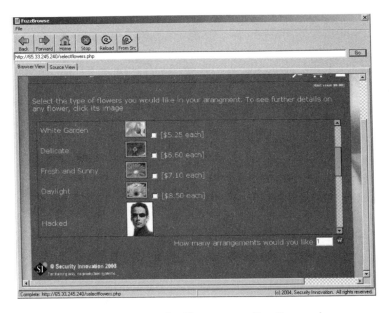

FIGURE 20.6 FuzzBrowse looking at an online flower shop.

order quantity. Figure 20.7 shows the source to this page. Highlighted is the JavaScript validation code that is checking to see that the user has put in a positive quantity less than 99. FuzzBrowse allows us to dynamically delete this JavaScript code and then feed a bad value into the page such as a quantity of –50. This value is then sent back to the server without the benefit of client-side validation, and we can test to make sure that it is indeed restricted on the server. FuzzBrowse can also be used to tamper with hidden fields and manipulate form restrictions.

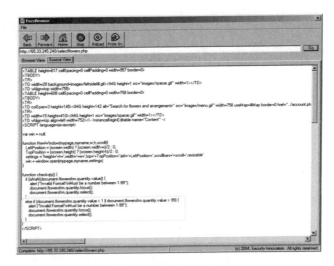

FIGURE 20.7 The source to the page from Figure 20.6.

FIXING THIS VULNERABILITY

Fixing these vulnerabilities boils down to extending minimal trust to the Web browser and user. For example, no one should ever be allowed to gain access to sensitive account information on a Web site based solely on an ID stored in a cookie. Cookies are great for tailoring Web content to a user or for remembering preferences but should not be the sole means of authentication to a system. The rule of thumb is, trust but verify. The same holds true for data that is obtained from a query string or a form. Any client-side checks on that data must be backed up with server-side validation.

Summary Sheet—Parameter Tampering, Cookie Poisoning, and Hidden Field Manipulation

Problem:

Data sent from a Web page back to a server is often trusted by the server. This trust might exist because of some client-side filtering done through JavaScript, restrictions placed on a field in HTML, or hard-coded hidden form fields. The issue is that any of these controls can be circumvented on the client, and data must be constrained on the server-side. Problems also occur when sensitive information is stored by the Web application in a user cookie. Two issues are apparent here. The first is that this data is potentially exposed to any user on the machine, and the second is that a malicious user can tamper with cookie data to potentially escalate privilege in the Web application or exploit the application's trust of this data.

Potential Impact:

Information disclosure, privilege escalation, and the potential to circumvent data restrictions that could lead to buffer overruns, SQL injection, and other vulnerabilities.

Habitat:

A wide variety of Web applications.

Tools You Need to Find It:

Most of the time these issues can be found by removing client-side validation controls and then manipulating data sent back to the server (long strings, escape characters, etc.). FuzzBrowse, a tool available from the author's Web site at *www.scottchase.net/FuzzBrowse/*, makes checking these types of issues easier.

How to Look for It:

For Web pages, look for the use of hidden form fields to store data. Try to reason about how this data is used and if any security relevant decisions are being made based on its value. Also look for client-side validation scripts that might check data type or length. Remove these controls, enter previously forbidden values, and see if these restrictions are also enforced on the server. It is also important to be keenly aware of what data is being stored in cookies. This data should be screened for sensitive user information as well as checked for values that are used to automatically authenticate a user or in some business relevant way (such as storing the price of an item for an e-commerce site). Be suspicious of all data that originates from a user's machine; it is under the user's control.

→

Symptoms of Failure:

This is a broad class of vulnerabilities and as a result symptoms of failure can vary wildly. When malicious data is entered into form fields, the results can range from a server application crash to an ODBC error message (indicating SQL injection potential). Once client-side validation has been removed, apply the techniques in the other Web chapters of this book and look for their respective failure symptoms.

Famous Failures/Exploits:

- CAN-2004-1209: VeriSign Payflow Link, when running with empty Accepted URL fields, does not properly verify the data in the hidden AMOUNT field, which allows remote attackers to modify the price of the items that they purchase.
- CAN-2003-0588: `admin.php` in Digi-News 1.1 allows remote attackers to bypass authentication via a cookie with the username set to the name of the administrator, which satisfies an improper condition in `admin.php` that does not require a correct password.
- CVE-2000-0720: `news.cgi` in GWScripts News Publisher does not properly authenticate requests to add an author to the author index, which allows remote attackers to add new authors by directly posting an HTTP request to the `new.cgi` program with an `addAuthor` parameter and setting the `Referer` to the `news.cgi` program.

REFERENCES

[Naraine04] Naraine, Ryan, "Barnes & Noble.com Fined for Customer Data Leak." *Internet News. www.internetnews.com/ec-news/article.php/3347761*, April 30, 2004.

21 SQL Injection Vulnerabilities

In This Chapter

- Exploiting Sites Through SQL Injection
- Finding This Vulnerability
- Fixing This Vulnerability
- References

Applications that operate on the Web often interact with a database to persistently store data. The way that Web application developers typically interface with a database is by constructing "queries" that manipulate or retrieve data stored in the database. For example, if an e-commerce application needs to store a user's credit card number, they typically retrieve the data from a Web form (filled out by the customer) and pass that data to some application or script running on the company's server. This application likely constructs a query to insert this data into a database. The dominant language that these database queries are written in is SQL, the Structured Query Language. (Table 21.1 describes some common commands in SQL. For a better reference, consult [Kline00].)

TABLE 21.1 Common SQL Commands and Special Characters

SQL Command	Description
condition AND *condition*	Used to intersect two conditions. Evaluates to true if both conditions are true.
condition OR *condition*	Evaluates to true if either condition is true.
EXEC master..*stored_proceedure*	Executes a stored procedure.
SELECT *columns* FROM *tables* WHERE *predicates*	Retrieves data from a table where certain conditions stated in *predicates* are met.
INSERT INTO *table* (*column-1, column-2, ... column-n*) VALUES (*value-1, value-2, ... value-n*)	Appends data (adds new record) to a table by putting the specified values into the listed columns.
UPDATE *table* SET *column* = *value* WHERE *predicates*	Replaces the data in the record(s) defined by *predicates* in the field specified by *column* with the included *value*.
' '	Single quotes are used to hold individual values.
;	Semicolon is used to separate SQL queries (for many database implementations).
—	Used for several databases (e.g., MS SQL Server) to treat everything after it as a comment.
*	Wildcard, meaning all.

Consider again the e-commerce application that needs to store a user's credit card information into a database. In SQL, the query is likely constructed using the data entered by the user. If the names of the form input fields are UserName and CreditCard for the user's name and credit card number, respectively, the SQL query to put that data into the Records table of the database might look something like this:

```
Query1 = "INSERT INTO Records (Name, CardNum) VALUES ('" & \
    Request.Form("UserName") & "','" & Request.Form("CreditCard") &
"')"
```

Here we are using the syntax of VBScript to construct our query. The "&" symbol concatenates our string, and Request.Form retrieves the value of the data entered into the form by the user. This construction typically takes place on the server, outside of the reach of an attacker. If, for example, we entered the name "Joe Smith" and a credit card number of "1234567890," the preceding query would end up looking like:

```
INSERT INTO Records (Name, CardNum) VALUES ('Joe Smith',
'1234567890')
```

Consider, though, another rather odd credit card number a user might decide to enter. If we were to enter a name, let's say "Fred Smith", with credit card number: "1'); EXEC xp_cmdshell 'del *.*'—", our query would now read:

```
INSERT INTO Records (Name, CardNum) VALUES ('Joe Smith', '1'); EXEC
xp_cmdshell 'del *.*'— ')
```

Those odd names, single quotes, and dashes that were entered are SQL commands that have now been *injected* into the query string. Let's take a look at what this new query does when executed. First, the double dash (—) in SQL signals that everything that follows it is a comment. Also, the semicolon concatenates two independent SQL queries. With this in mind, we are now executing two separate queries. The first simply inserts a new record into the Records table with name "Joe Smith" and credit card number "1". The second, however,

```
EXEC master..xp_cmdshell 'del *.*'
```

executes the xp_cmdshell stored procedure, which in turn can execute arbitrary commands on the server. In this case, our second query deletes all files in the directory in which the database server's key binaries are located. By default, in Windows 2000 and XP running SQL Server, this is the \Program Files\ folder. We could do much more devious things than this, however, like share out a directory or spawn a remote shell. Even without using stored procedures, though, the fact that we can directly manipulate the query string on the server is troubling. For queries that return data to the user or check authentication, we can manipulate the query string to access arbitrary records in a database or bypass authentication. Table 21.2 describes some common table names that can be manipulated on different database servers.

TABLE 21.2 System Table Names in Some Common Database Servers

Microsoft SQL Server	Microsoft Access Server	Oracle
sysobjects	MSysObjects	SYS.ALL_TABLES
syscolumns	MSysACEs	
SYS.USER_TAB_COLUMNS		
	MSysQueries	SYS.USER_OBJECTS
	MSysRelationships	SYS.TAB
		SYS.USER_TABLES
		SYS.USER_VIEWS
SYS.USER_CONSTRAINTS		
		SYS.USER_TRIGGERS
		SYS.USER_CATALOG

The key problem here is that user input is not validated on the *server* before it is used in a query. Many Web developers rely on client-side validation to protect against escape characters being entered by the user. Typically, a Web form is validated using JavaScript where characters such as single quotes are either stripped or replaced using client-side routines before this data is passed back to the server. This is the method of choice for many developers because the bulk of the work is done on the client, thus increasing server performance. The following is a typical example of JavaScript client-side validation of a form.

```
<SCRIPT>
checkval=new RegExp("[\-\'\;]");

function validate(){
if (checkval.test(form1.Acct.value)){
    alert("Account names and passwords should not contain special
characters");
    event.returnValue=false;
    }
if (checkval.test(form1.Pin.value)){
alert("Account names and passwords should only contain numbers
    and letters");
    event.returnValue=false;
    }
```

```
}
.
.
.
<FORM name="form1" action="process.asp" method="post"
onsubmit="validate();">
.
.
.
```

If illegal characters are included in either of the input fields, the user is presented with a warning and is forced to replace the offending characters. The critical flaw here is that this validation against SQL injection is done on the *client*. A malicious user can easily circumvent this protection by saving the Web page and commenting out the call to the validation routine. The attacker would then need to include the full URL to the page that processes this data and simply reload saved page in a browser. The modified HTML source is shown here:

```
<FORM name="form1" action="http://www.si-hackedbank.com/
process.asp">
```

Threats such as these can be avoided if user data is stripped of all SQL commands and delimiters on the *server side* before it is used to construct a database query.

A related problem that many Web sites have is the disclosure of information through Open Database Connectivity Protocol (ODBC) error messages. Figure 21.1 shows a typical example. Once an attacker sees a screen such as this, he knows it is likely that a SQL injection vulnerability exists. From this error message an attacker gains a significant amount of information:

- On the client or the server is some user-string validation or error-handling routine that is either missing or incorrectly written. Conclusion: this weakness can be exploited to change the SQL query on the server.
- Database errors can reveal an amazing amount of information about the query string that is being executed. By changing input to the form we can eventually discover most or all of the query string on the server. The query string is likely to include key table names and field names that can then be leveraged to mount more complex and insidious SQL injection attacks.
- Such errors reveal which database server is running on the back end and thus make an attacker's job easier by narrowing the techniques he uses. In this case, we see that Microsoft's SQL server is processing user data.

316 The Software Vulnerability Guide

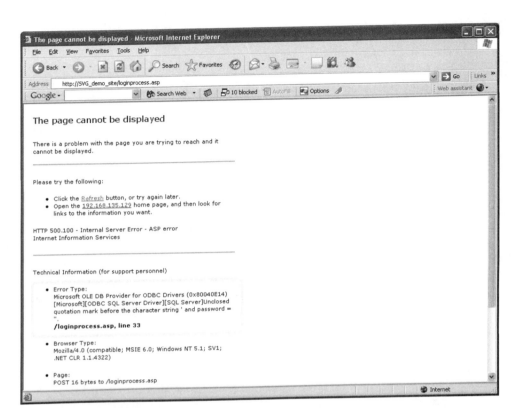

FIGURE 21.1 ODBC error messages like this one reveal information that is helpful to attackers.

EXPLOITING SITES THROUGH SQL INJECTION

Typically, an attacker's first clue that a Web site is vulnerable to SQL injection is an error message being thrown. When a query is crafted so that a set of data is returned to the user on a subsequent page, then an attacker can typically manipulate the SQL query so that additional information can be returned through the browser. If an ODBC error message is returned indicating that the server is running Microsoft's SQL, then an attacker is likely to attempt to run a stored procedure (discussed earlier in the chapter). For example, consider the Web site shown in Figure 21.2 that demonstrates a typical username and password field.

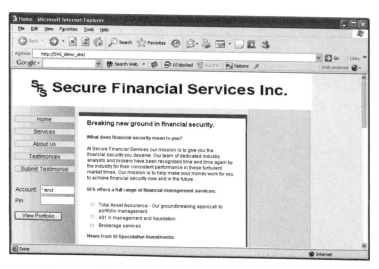

FIGURE 21.2 Information from login screens is usually validated against information stored in a database.

Entering the string ' and into the username field results in the error message shown in Figure 21.1. An attacker could now proceed in several ways. By default, SQL Server 2000 allows the execution of many stored procedures, and therefore, one option is to execute one of these by appending an additional command to a SQL statement. By entering the following string in the "Account" field, a user could potentially do a directory listing of the C: drive and save it to a text file in a typically accessible location.

```
'; EXEC master.dbo.xp_cmdshell 'cmd.exe /c dir c:\ > C:\inetpub\wwwroot\dir_c.txt'--
```

Execution of the `xp_cmdshell` stored procedure requires that the Web application is connecting to the database server as the "sa" user, which has full control over the database. This is a common yet incredibly insecure practice for Web developers because it violates the principal of least privilege: connect with the minimum privileges necessary to accomplish the task. All that would be required to view this output is to browse to the file *http://www.si-hackedbank.com/dir_c.txt*. If stored procedures are not enabled or if the target were a different database server, then the next step would be to do some reconnaissance on the tables/fields of interest. Table 21.2 shows the names of the tables that typically contain names and associated columns for various database servers. To display these names, we must union the current query with one designed to return the information necessary.

To continue with the example application, if a legitimate username/pin is entered, then all of the records are displayed from that user (see Figure 21.3).

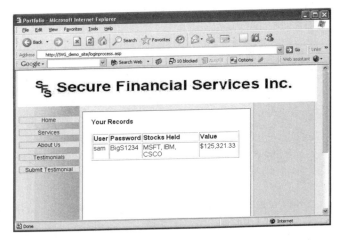

FIGURE 21.3 When a correct username and password is entered, this Web application shows the user's account information.

Now, instead, we want this page to return some data from the `syscolumns` table, which contains information about other tables in the database. The names of the tables are in the `name` column of the `sysobjects` table. To execute our query, we must use the UNION command to join two SELECT commands: the one present on the page originally and our new query on `syscolumns`. To use the UNION operator, we have to select the same number and type of elements in each query. For example, if we enter the string:

```
' union all select name,0,0,0,0 from sysobjects--
```

into the "Account" field, the error message in Figure 21.4 is returned.

We can thus determine both the number and type of parameters in the original query. An attacker might be interested in many things. One of these is a list of the tables and whether or not they were created by the user. The `name` column of `sysobjects` contains table names, and the `xtype` column reveals which user created the tables. In the running example, we can list the tables using the following input in the "Account" field:

```
' union all select name,xtype,0,0 from sysobjects--
```

SQL Injection Vulnerabilities

FIGURE 21.4 This error message informs us that the number of columns we entered needs to be different. Through trial and error (usually within just a few attempts), these error messages can reveal the combinations necessary to launch our attack.

The result (Figure 21.5) is a list of all tables and their corresponding type. Of particular interest is likely to be the Records table because it is the only one created by a user (indicated by an xtype of U).

By applying techniques similar to this, an attacker could now view all data contained in the database. Using the INSERT and UPDATE commands, an attacker can also make alterations to the data.

FINDING THIS VULNERABILITY

One of the most basic techniques for finding SQL injection vulnerabilities is to enter SQL escape characters into input fields. A single quote is a good start. Next, look for an error message to be returned. Messages like the one shown in Figure 21.4 mean that it's highly likely that an attacker can manipulate the database or the entire system by altering the SQL query.

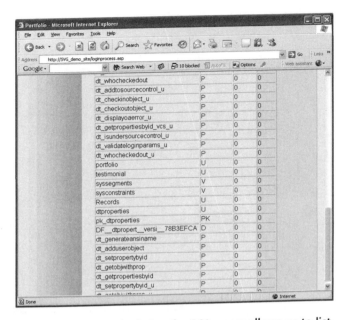

FIGURE 21.5 Manipulating the SQL query allows us to list all tables in the database. This is usually all the information we need to launch an effective attack.

The listing of Index.html shows the vulnerable implementation of client-validated user inputs, which are then used directly in SQL queries shown in Process.asp. Note that no validation of user input is performed on the server side (in Process.asp). Such implementations can easily lead to server vulnerabilities because a user can bypass client validation and thus manipulate the server-executed query (see sidebar "Exploiting Sites Through SQL Injection" for details).

Index.html

```
<HTML>

<SCRIPT>
checkval=new RegExp("[\-\'\;]");

function validate(){
if (checkval.test(form1.Acct.value)){
    alert("Account names and passwords should only contain numbers
    and letters");
    event.returnValue=false;
```

SQL Injection Vulnerabilities

```
            }
        if (checkval.test(form1.Pin.value)){
            alert("Account names and passwords should only contain numbers
            and letters");
            event.returnValue=false;
            }
        }
        .
        .
        .
        <FORM name="form1" action="process.asp" method="post"
        onsubmit="validate();">
        .
        .
        .
            <TD>Account</TD>
            <TD><INPUT type="text" name="Acct" size="20"></TD>
            </TR><TR>
            <TD>Pin #</TD>
            <TD><INPUT type="password" name="Pin" size="20"></TD>
        .
        .
        .
        </FORM>
        </HTML>
```

Process.asp

```
        <%@ LANGUAGE = VBScript %>
        <% Option Explicit %>

        <%
        .
        .
        .
            QueryName = "SELECT * FROM Records WHERE Username = '"
            QueryName = QueryName & Request.Form("Acct") & "' and Pin = '" &
            Request.Form("Pin") & "'"

            Set oRs = oConn.Execute(QueryName)
        %> Your Records</p>

        <TABLE border = 1>
```

```
<%
    Do while (Not oRs.eof) %>
        <tr>
        <% For Index=0 to (oRs.fields.count-1) %>
        <TD VAlign=top><% = oRs(Index)%> </TD>
        <% Next %>
        </tr>
        <% oRs.MoveNext
    Loop
%>
.
.
.
```

FIXING THIS VULNERABILITY

Index.html executes client-side validation and returns error messages to the user. This is a good feature, but it is certainly not sufficient to protect this Web application. Regardless of client-side validation, data must be validated on the server before it is used in a query. In the following we show one example of how to implement server-side validation in Process.asp.

Process.asp

```
<%@ LANGUAGE = VBScript %>
<% Option Explicit %>

<%
.
.
.
'Create a parameterized command and set the active connection to the
'connection established

Set sqlcmd = CreateObject("ADODB.Command")
sqlcmd.CommandText ="SELECT * FROM Records WHERE Username=? and Pin =?"
sqlcmd.ActiveConnection = conn
sqlcmd.CommandType = 1
sqlcmd.Prepared = true


'The CreateParameter function takes five parameters: name, type,
'direction, size, value
```

```
'type 200 is varchar string
'direction 1 is input parameter only
'for Username size is 32 and for Pin size is 8

Set sqlparam1 = sqlcmd.CreateParameter("Username", 200, 1, 32, "")
cmd.Parameters.Append sqlparam1
sqlparam1.Value = Request.Form("Acct")

Set sqlparam2 = sqlcmd.CreateParameter("Pin", 200, 1, 8, "")
cmd.Parameters.Append sqlparam2
sqlparam2.Value = Request.Form("Pin")

Set oRs = cmd.Execute

   .
   .
   .
%>
```

The fixed example shows validation of form data on the server. Server-side validation is essential for user-supplied data through the Web. By stripping escape characters from user data, we can ensure that a user cannot manipulate the executed SQL query.

Summary Sheet—SQL Injection

Problem:

Web applications can be vulnerable to a malicious user crafting input that gets executed on the server. One instance of this is an attacker entering Structured Query Language (SQL) commands into input fields, and then this data being used directly on the server by a Web application to construct a database query. The result could be an attacker's gaining control over the database and possibly the server. Care should be taken to validate user input on the server side before user data is used.

Potential Impact:

End-user control over database information and possibly the database/Web server.

Habitat:

Web applications that interface with a database. General applications that use a database.

Tools You Need to Find It:

Web browser

How to Look for It:

The idea is to get escape characters in data that is used by the Web application to construct a database query. Queries are constructed usually to retrieve data from a database, add data to a database, or change data in a database. Most of the time, queries are created based on data entered by a user on a previous page. To find this vulnerability, you need to force the Web application to construct its query using user data in a way that can change how that query behaves. For Web applications, data is passed from page to page or from a Web page to the Web server using two primary methods: POST and GET. The GET method shows user data in the URL, whereas POST passes data through the HTTP header, which is not readily visible through a browser. Both methods are vulnerable and can be tested by entering SQL escape characters and watching for database error messages.

Symptoms of Failure:

If you are successful at getting SQL delimiters into queries on the server, a good indicator of application security failure is an error message being returned through the browser that complains about bad syntax. Figure 21.1 shows an example of what this might look like.

Famous Failures/Exploits:

- CVE-2001-1053: `AdLogin.pm` in AdCycle 1.15 and earlier allows remote attackers to bypass authentication and gain privileges by injecting SQL code in the `$password` argument.
- CVE-2002-0287: pforum 1.14 and earlier does not explicitly enable PHP magic quotes, which allows remote attackers to bypass authentication and gain administrator privileges via an SQL injection attack when the PHP server is not configured to use magic quotes by default.

REFERENCES

[Kline00] Kline, Kevin and Kline, Daniel. *SQL in a Nutshell*. O'Reilly and Assoc., 2000.

22

Additional Browser Security Issues

In This Chapter

- The Domain Security Model
- Unsafe ActiveX Controls
- Spoofing of URLs in the Browser
- MIME Type Spoofing
- Uncommon URL Schemes
- Browser Helper Objects

As the Web browser has transformed from a simple hypertext tool into an extensive application development platform, a number of features that provide functionality outside the traditional "Web page" experience have been added. The browser has been augmented even more as integration with the operating system and windowing system has increased and as developers begin to use the browser model even for purely client-side applications. This superset functionality, which still contains HTML and its accessory technologies as its base, presents its own unique set of security issues.

In the previous chapters, we talked about vulnerabilities that are common to most Web browsers and application programming paradigms. This chapter deals with issues that are specific to a subset of browsers, especially Microsoft Internet Explorer. Because that one application (IE) has, at the time of this writing, more than 90 percent of the market share for end user Web browsing, its special issues are worth noting.

Since its original decision to integrate the Internet Explorer browser with Windows 95, Microsoft has progressively increased interoperability between the browser and the operating system and desktop programming paradigm. This has lead to a number of developments that make it easier for programmers to program using the Web browser as a base:

- Through ActiveX controls accessible through JavaScript and VBScript, browser-based applications can invoke any COM-compatible objects or applications present on the client system. This has several practical uses. First, Web programmers can use these controls to launch common applications, such as Word or Excel within the browser frame, creating a less "cobbled" appearance and user experience when a Web application needs to interact with an object associated with one of these applications. Because developers can write their own controls, it also means that browser functionality can be extended by pushing an ActiveX control down to the user's machine and then invoking the functionality within that control. Many common Web extensions such as Macromedia Flash work in this fashion.
- Because the MSHTML object has an exposed COM interface, application developers who want to integrate the browser into their desktop applications can do this. As a result, many applications incorporate the browser into their application for file browsing, text viewing, online help, and easy to program dynamic forms.
- Microsoft has added a number of proprietary URL monikers (for protocols other than *http://*, *ftp://*, and *mailto://*) to accommodate online help, internal browser functionality, and browser extensions. These monikers are generally supported in the same fashion as the major ones within the browser.
- Browser Helper Objects, designed to extend the functionality of Internet Explorer for legitimate purposes, have been co-opted by spyware producers as a means of controlling the victims' browser and stealing their personal information.

Each of these extensions, together with numerous enhancements made to deal with specific vulnerabilities, has made security within the modern browser decidedly more complicated. In this chapter, we explore each of these issues in more detail.

THE DOMAIN SECURITY MODEL

One of the conveniences of browser programming in IE is its Document Object Model (DOM). The DOM is a set of JavaScript objects that manipulate the browser and the pages loaded within it; it is generally possible to modify any attribute of a page through the object model, as well as open, close, move, and resize windows; prompt for input or file selection; and pop up message boxes. To protect against a malicious page from one domain manipulating the objects of a page within another domain, IE implements a *domain security model*. It is called the domain security model because the security settings are determined based on the domain name encoded in the URL. The domain security model also protects against a variety of potentially unsafe behaviors:

- It prevents a malicious Web site from using "local" URLs, such as those with the *file://* moniker to manipulate local file system objects.
- It limits the use of Java, JavaScript and ActiveX controls in certain domains based on the user's security settings.
- It forces prompting for download of unsafe file types based on the user's settings.

The domain security model categorizes a URL as being in one of four "zones." Figure 22.1 shows the settings dialog for these security zones within Internet Explorer.

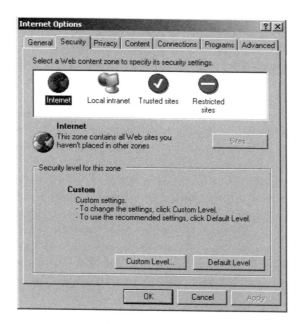

FIGURE 22.1 The settings dialog for these security zones within Internet Explorer.

- The "Local intranet" zone includes all URLs that reference files located on the local machine or accessed through Windows networking within the same domain as the local machine. When the browser views a page in the local intranet zone, its chief responsibility is to prevent the leakage of information accessible through that zone back to the Internet. That way, a malicious script cannot be used to steal user's data off the drive.
- The "Internet" zone is used for ordinary URLs that point to pages in other Internet domains. The vast majority of sites are in the "Internet" zone. The key security concerns of the browser in the Internet zone are to prevent access to functionality that can read and retransmit contents of the local filesystem, and to prevent manipulation of DOM objects across different domains.
- The "Trusted sites" zone includes sites the user has given specific trust permissions to bypass some of the domain security settings. These are typically either intranet sites that don't conform to the naming policy used by the domain security model that the user wants to give additional privileges to (for example, if our own domain name were *XYZLifeInsurance.com*, we might assign URLs belonging to our parent company, *XYZHoldings.com* the same behavior as intranet sites, even though IE would see them as separate domains) or sites that are used for automatic updates.
- On the other hand, some sites might warrant extra precaution. IE gives the user the ability to place these sites in a "Restricted sites" zone. The user can lock down the security settings at a restricted site, for example, to prevent all Java, JavaScript, and ActiveX, without imposing these restrictions on ordinary Internet zone sites. The "Restricted sites" zone is typically used for untrustworthy sites that a user needs to access anyway. For example, we use the "Restricted sites" zone for some of the sites we use to download exploits and shellcode for examination and testing. For some reason, we just don't trust the people who post exploits on the Internet with the ability to execute scripts on our machine.

A number of vulnerabilities are caused by Internet Explorer's inappropriately applying these security settings, or by attackers finding a way around them. An attacker can often use these vulnerabilities to send a malicious e-mail or URL to a victim, which when opened, facilitates a file download, redirection to a malicious Web site, or other such attack. The security researchers Liu Die Yu and Georgi Guninski have discovered a number of these vulnerabilities, which are published on their Web sites, *http://umbrella.name/* (Die Yu) and *www.guninski.com* (Guninski).

UNSAFE ACTIVEX CONTROLS

As we mentioned previously, the browser has the capability to invoke any ActiveX control installed on the system that has been marked "Safe for Scripting" or "Safe for Initialization." An ActiveX control, in turn, can perform any operation that is allowable on the local computer, because it is just an executable program. Frequently, vendors distribute ActiveX controls that are marked "Safe for Scripting" but have the ability to programmatically access files on the local filesystem or contain buffer overflow vulnerabilities. These controls essentially open a back door for JavaScript running in the "Internet" zone to access the local filesystem and execute arbitrary code. Controls should not be marked "Safe for Scripting" unless they do not access the local filesystem in an arbitrary fashion and have been thoroughly tested against buffer overflows and other kinds of input attacks.

SPOOFING OF URLS IN THE BROWSER

Remember how we said that the Document Object Model can be used to manipulate portions of the browser user interface as well as the documents within it? This has led to a number of creative ways of "spoofing" URLs—making it appear as though the URL in the address bar at the top of the page or in the status bar at the bottom of the page is different from the URL that is actually loaded. An attacker who wants to trick a user into entering personal information uses URL spoofing to make his malicious site appear as though it is the legitimate site.

IE has contained a number of bugs over the years, many now fixed, that have accommodated URL spoofing. The techniques include:

- Embedding null characters or delimiters in the malicious URL. Older versions of the browser would navigate to a URL such as *http://www.anyonesbank.com%01%00@hackerbank.com* as though *www.anyonesbank.com%01%00@* were just a username at *hackerbank.com*, but the user interface controls, after converting the *%01%00* to a Unicode null character, truncate the string to *www.anyonesbank.com*. The result is that an attacker could fool an unsuspecting user into thinking that *hackerbank.com* was really *www.anyonesbank.com*.
- Using non-printable characters or spaces in URLs. A malicious Web site might contain a URL that is composed of the "legitimate" site name (such as *www.ayonesbank.com*) followed by many nonprinting characters and then the real domain name (*hackerbank.com*).) In this case the browser UI displays the URL, but it has so many spaces or nonprinting characters that the real domain name is displayed outside the limits of the UI control.
- Previously, a JavaScript could obtain a handle to the Document Object Model of a page in a remote domain if that page was embedded in a frame within the browser. In this case, the attacker's site could actually load the legitimate site in a frame, especially in a pop-up window that did not have a URL bar. JavaScript within the attacker's page could then read the values of form fields, such as usernames and passwords, out of the child frame.

- Overloading of exit events. Overloading the exit event with code that causes an infinite loop can prevent correct updating of the URL in the address bar. This vulnerability existed in previous versions of the Opera Web browser.
- Topographical attacks. A number of clever attacks involve creating images, "chromeless" (borderless) windows, or Java applets that physically conceal the real address bar with a fictitious one. These attacks work because the screen resolutions and browser geometry of many users are similar.

MIME TYPE SPOOFING

Multipurpose Internet Mail Extensions (MIME) is a standard means of encoding attachments in e-mails so that they can be decoded and delivered to the proper application. The MIME type is a string describing the data as text or binary, as well as its contents. HTTP also uses MIME to transfer content contained in Web pages. For example, the MIME type of an ordinary Web page is "text/html." This type name is transmitted along with the data in an HTTP reply.

Internet Explorer determines the default action for a file (displaying in browser, saving to disk, launching a handling application, or launching the file as an executable) based on a combination of MIME type and file extension. MIME spoofing occurs when a malicious Web site tries to trick the browser into opening a file it shouldn't by giving it a bogus MIME type. The attacker would usually try to disguise a dangerous type, such as an .EXE, as a more innocuous type that can be opened automatically in the browser. MS01-20 described a MIME spoofing vulnerability in which IE incorrectly handled certain unusual MIME types; if an attacker created an HTML e-mail containing one of these types, IE would launch the attachment as though it were an executable without first prompting the user.

UNCOMMON URL SCHEMES

Internet Explorer supports a number of protocols and URL schemes beyond HTTP. Sometimes these schemes, intended for use only by Microsoft or only on the local machines, have unexpected consequences. Like ActiveX controls, they sometimes provide back doors that an attacker's Web page can use to gain access to the local filesystem or execute arbitrary commands.

Some of the more unusual ones include:

- **shell:** This protocol is used by IE to load resources based on special "shell" protocol names. A bug in the way IE handled security zones with this protocol meant an attacker could execute malicious scripts in the local zone, where he

could access arbitrary files. This bug has recently appeared in the Mozilla browser as well.
- **mk:** This protocol is used by IE to designate that the URL uses a protocol moniker that is not in the default set and must be accessed by name. URL parsing bugs with the mk: protocol could sometimes be used to execute malicious scripts in the local zone.
- **hcp:** This protocol is used by Microsoft Help and Support Center, the new default Help application for the Windows shell. hcp: is implemented with JavaScripts stored in resources within the application; some of these JavaScripts contained errors that could allow an attacker who supplied a malicious URL to execute arbitrary programs or delete files.
- **res:** IE uses the res: protocol to access Windows resources within DLLs and applications. Previously, some cross-site scripting vulnerabilities existed in local JavaScripts accessible via res:, which an attacker could exploit.

All of the vulnerabilities described have been fixed previously or will be fixed via Windows XP Service Pack 2. However, as IE has not removed support for these protocols, and several more are undocumented, additional vulnerabilities might be identified in the future.

BROWSER HELPER OBJECTS

Browser Helper Objects are used to extend the in-browser functionality of Internet Explorer in a way that works across all pages. (Java, JavaScript, and ActiveX can work only within the context of a single page or set of pages.) The Google Toolbar Helper Object, for example, adds a search toolbar, context menus, and pop-up advertisement blocker to IE. Other BHOs have more nefarious uses; many spyware creators use BHOs to record all of the URLs a victim accesses, to manipulate search results, or to redirect error pages to advertisements. Prior to Service Pack 2, a user did not have an easy way to see or remove Browser Helper Objects installed on his computer.

Summary Sheet—Additional Browser Security Issues
Problem:

Modern Web browsers such as Internet Explorer provide additional functionality beyond the ability to browse hypertext pages. Some of this functionality is well understood, but a number of lesser known and undocumented features exist. Attackers with knowledge of some of these issues can use them to exploit a user's machine.

Potential Impact:

The impact of browser security issues depends on the individual issue. In some circumstances, the risk is confined to enumerating (but not accessing) the files on a user's local filesystem or stealing a cookie. On the other hand, in some cases these vulnerabilities allow execution of arbitrary code or theft of a user's personal information.

Habitat:

Later versions of the Web browser, especially those that integrate tightly with the operating system, are where these vulnerabilities are found.

Tools You Need to Find It:

A registry inspection tool such as regedit (supplied with Windows) or Regmon (from *Sysinternals.com*) can be used to find some of these issues, including potentially unsafe ActiveX controls and Browser Helper Objects. Others must be found by hand or by trial and error.

How to Look for It:

The `HKEY_LOCAL_MACHINE\SOFTWARE\Microsoft\Internet Explorer\ActiveX Compatibility\` registry branch contains the compatibility flags for `ActiveX controls`.
A list of Browser Helper Objects installed on a machine can be found in `HKEY_LOCAL_MACHINE\SOFTWARE\Microsoft\Windows\Current Version\Explorer\Browser Helper Objects`.

Symptoms of Failure:

Look for controls without any obvious purpose, controls with no names, or controls with known vulnerabilities.

Famous Failures/Exploits:

- The "%01" bug in Internet Explorer 5 allowed an attacker to bypass domain security.
- Liu Die Yu (*http://umbrella.name/*) and Georgi Guniski (*www.guninski.com/*) have a wealth of information about browser security issues on their Web site.

Part VII Conclusion

23 Conclusion

In This Chapter

- Learning from Vulnerabilities
- Where to Go Next
- References

Software security has become a critical issue for every business and nearly every home user. Security is a complicated issue, though, and spans software, configuration management, network perimeter security, people, and processes. While non-software issues are important to manage, a recent survey by the leading technology analyst firm Gartner estimates that 70 percent of vulnerabilities in a system exist at the application layer [Pescatore03]. It is clear then that the traditional paradigm of fortifying the network perimeter for security is inadequate and that there is a direct correlation between spending on network defenses and information security. We are now beginning to understand that the point of diminishing returns for network defenses is smaller than originally thought, and the types of attacks that can be stopped by current network defenses is small in comparison to the vulnerabilities that can exist on a system because of insecure applications.

This is becoming a severe issue for corporate IT managers as the volume of attempted intrusions continues to rise. Figure 23.1 shows how the number of security incidents reported to the CERT Coordination Center has increased over the last several years.

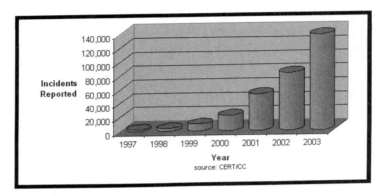

FIGURE 23.1 The number of security incidents reported to the CERT Coordination Center has increased over the last several years.

With the limitations of network defenses in mind, the need to produce applications that are more resilient to attacks is acute, and the demand for developers and testers who are security savvy continues to increase. This book presents a roadmap for software security vulnerabilities, one that is designed to help developers, testers, and managers better understand the enemy. But we face significant obstacles. The software industry continues to struggle with the functionality/security tradeoff, and it is still clear that even if we build secure software we still don't have a handle on the human element of security. This is clear by the increased popularity of phishing attacks that impersonate a legitimate business and goad users into disclosing their usernames and passwords to an attacker.

Consider Figure 23.2, an e-mail that purports to be from a major bank, alarming readers and telling them that they must click on the included link to "secure" their account. Although the URL looks legitimate, the link actually transports the user to the attacker's site, which is a complete mock-up of the legitimate bank's real site, as shown in Figure 23.3. When the unsuspecting user clicks on the link, he is transported to the mock-up, which asks for login credentials. This information is then used by the attacker to compromise that user's account on the real banking site.

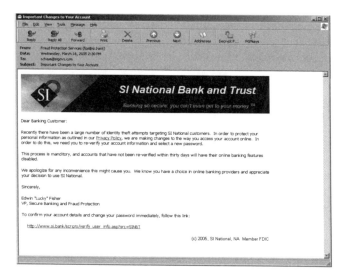

FIGURE 23.2 E-mail of a fictitious "phishing" attack.

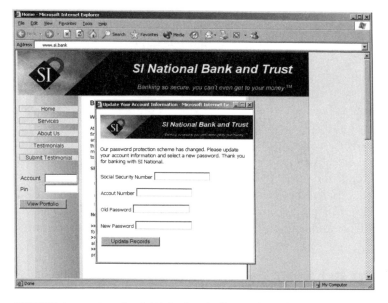

FIGURE 23.3 Attacker's Web site similar to that of our fictitious bank.

These types of attacks illustrate the limits of secure coding. While both the e-mail application and the Web browser might be relatively secure, they must be usable. This usability means that an attacker can exploit user naïveté and force them to do something inherently dangerous. As developers and testers, we must be the ones who are constantly evaluating the security, functionality, and security continuum and make best efforts to rid our applications of known security errors.

LEARNING FROM VULNERABILITIES

We can learn a tremendous amount by looking at the vulnerabilities in our applications and the mistakes of others. At our company, whenever we finish a development or testing project, we bring everyone into a room for a day or several days and go over the bugs we caught late in the process. Thoughtfully studying security mistakes can be one of the best ways to detect and prevent those issues in future projects. These evaluations help refine the design, development, and testing process to ensure that vulnerabilities are addressed earlier in the software development lifecycle. They can also help a team highlight holes in its quality assurance process and sharpen techniques to find certain classes of security failures. Such evaluations are best done soon after the testing project ends, when bugs are still fresh in the testers' minds. These studies are also extremely valuable when performed periodically for released or deployed products, particularly when bugs are uncovered in the field.

Many organizations have also found it valuable to study their competitors' bugs to ensure that they have methods in place to keep similar bugs out of current projects. Some fantastic free resources can be found on the Internet to get examples of vulnerabilities in deployed applications. One of our favorites is BugTraq. The BugTraq mailing list (available at *www.securityfocus.com*) provides a continuous source of field-reported security vulnerabilities. We encourage you to subscribe to the list and spend a short period of time looking at the vulnerabilities that pass its way. Another great source is the Common Vulnerabilities and Exposures repository sponsored by the U.S. government and run by Mitre (*cve.mitre.org*). CVE is a great reference for a variety of software vulnerabilities and is fairly easy to search by keyword. Finally, the Computer Emergency Response Team (CERT) site at Carnegie-Mellon is a good source of not just software vulnerabilities but of security related incidents and statistics as well. CERT's Web site is *www.cert.org/*.

WHERE TO GO NEXT

Security knowledge is dynamic, and to keep ahead of attackers requires a constant knowledge refresher. Besides staying current on the latest vulnerabilities, you can do a few other things to improve your knowledge of computer security:

- Attend a conference for security-minded professionals. The most famous of these is DEF CON, which bills itself as the "largest underground hacking event in the world." DEF CON is held annually in Las Vegas and is popular with both security professionals and "black hat" types. DEF CON is somewhat more focused on the *culture* of computer security, including the social habits of traditional black hat hackers. Black Hat, which is held five times annually in Seattle, Las Vegas, Washington D.C., and Japan, offers somewhat more technical presentations. Many of these are "cutting edge" and so might not be suitable to the security beginner, but others are suitable for everyone. The RSA Conference, which also hosts events in the United States, Europe, and Asia, is the major trade event for security vendors. While some of the tracts here are related to security from an IT or business perspective, many are also good for developers.
- Attend an application security training course. Foundstone, based in Mission Viejo, California offers "Ultimate Hacking" and "Ultimate Web Hacking," as well as a variety of other courses to developers. Security Innovation in Melbourne, Florida offers "How to Break Software Security," a two-day course that covers the techniques described in this book. Black Hat, the company that sponsors the conference by the same name, also offers application security training.
- Review the trade literature. *Dr. Dobbs Journal*, the largest circulation magazine for developers, devotes one issue annually to security. The Institute for Electrical and Electronics Engineers (IEEE), a major professional society for computer scientists and engineers, has created a new magazine, *Security and Privacy*, which is entirely devoted to information security. *Information Security* magazine covers security issues including the latest vulnerabilities.
- Get a security certification. The Certified Information Systems Security Professional (CISSP) is the most prestigious of these. This certification is not just for IT professionals and is widely respected in the industry. Information about it can be found at *www.cissp.com*.

Whatever you do, don't read this book and then give up on security. Ultimately, it is up to individual developers to improve the security landscape. Happy hunting!

REFERENCES

[Pescatore03] Pescatore, Joe. Gartner. "CIO Alert—Follow Gartner's Guidelines for Updating Security on Internet Servers, Reduce Risks." February 2003.

Appendix A: About the CD-ROM

The CD-ROM included with *The Software Vulnerability Guide* contains the code and projects from the various examples found in the book. In addition, it includes a number of open-source security testing tools discussed in Chapter 3, "Some Useful Tools."

CD-ROM FOLDERS

Companion Tools: This folder contains the open source security testing tools described in Chapter 3 and used throughout the book. The individual tools included are:

- Libnet (*www.packetfactory.net/libnet*)
- Ethereal (*www.ethereal.com*)
- Ettercap (*ettercap.sourceforge.net*)
- John the Ripper (*www.openwall.com/john*)
- Nemesis (*nemesis.sourceforge.net*)
- Nessus (*www.nessus.org*)
- Nikto (*www.cirt.net/code/nikto.shtml*)
- Nmap (*www.insecure.org/nmap*)
- RATS (*www.securesoftware.com*)
- SATAN (*www.fish.com/satan*)
- Tcpdump (*www.tcpdump.org*)

Source Examples: Contains the source examples and project files included in each of the chapters.

Images: The images from each chapter in the book.

OVERALL SYSTEM REQUIREMENTS

- Windows 2000 or Windows XP (Windows Examples and Tools)
- Linux for X86-family machines, Kernel Version 2.2 or Higher (Linux Examples and Tools)
- Pentium II Processor or greater
- CD-ROM drive
- Hard drive
- 128 MBs of RAM (Minimum 256 recommended for Windows)
- Microsoft Visual Studio 6.0 or higher (Windows)
- *GNU GCC 2.8* or higher (Linux)
- 50 MBs of hard drive space for the code examples and tools.

You will need a compiler or development environment for Windows to compile the code examples. Some examples have an associated project file; in order to use these you will need Microsoft Visual Studio 6.0 or higher. In addition, you will need Visual Basic 6.0 to compile the Visual Basic and ASP examples. Many of the tools for Linux are provided in source form; you will need GCC in order to compile these.

INSTALLATION INSTRUCTIONS

The source examples require no installation; simply open the project file (.vbp) or individual source files in Visual Studio. Each of the companion tools should be installed separately by following the instructions for that tool. The complete instructions for each tool are included on the CD-ROM.

Appendix B: Open Source Software Licenses

GNU GENERAL PUBLIC LICENSE

GNU General Public License
Version 2, June 1991
Copyright (C) 1989, 1991 Free Software Foundation, Inc.
675 Mass Ave, Cambridge, MA 02139, USA

Everyone is permitted to copy and distribute verbatim copies of this license document, but changing it is not allowed.

Preamble

The licenses for most software are designed to take away your freedom to share and change it. By contrast, the GNU General Public License is intended to guarantee your freedom to share and change free software—to make sure the software is free for all its users. This General Public License applies to most of the Free Software Foundation's software and to any other program whose authors commit to using it. (Some other Free Software Foundation software is covered by the GNU Library General Public License instead.) You can apply it to your programs, too.

When we speak of free software, we are referring to freedom, not price. Our General Public Licenses are designed to make sure that you have the freedom to distribute copies of free software (and charge for this service if you wish), that you receive source code or can get it if you want it, that you can change the software or use pieces of it in new free programs; and that you know you can do these things.

To protect your rights, we need to make restrictions that forbid anyone to deny you these rights or to ask you to surrender the rights. These restrictions translate to certain responsibilities for you if you distribute copies of the software, or if you modify it.

For example, if you distribute copies of such a program, whether gratis or for a fee, you must give the recipients all the rights that you have. You must make sure that they, too, receive or can get the source code. And you must show them these terms so they know their rights.

We protect your rights with two steps: (1) copyright the software, and (2) offer you this license which gives you legal permission to copy, distribute and/or modify the software.

Also, for each author's protection and ours, we want to make certain that everyone understands that there is no warranty for this free software. If the software is modified by someone else and passed on, we want its recipients to know that what they have is not the original, so that any problems introduced by others will not reflect on the original authors' reputations.

Finally, any free program is threatened constantly by software patents. We wish to avoid the danger that redistributors of a free program will individually obtain patent licenses, in effect making the program proprietary. To prevent this, we have made it clear that any patent must be licensed for everyone's free use or not licensed at all.

The precise terms and conditions for copying, distribution and modification follow.

GNU GENERAL PUBLIC LICENSE
TERMS AND CONDITIONS FOR COPYING, DISTRIBUTION AND MODIFICATION

0. This License applies to any program or other work which contains a notice placed by the copyright holder saying it may be distributed under the terms of this General Public License. The "Program," below, refers to any such program or work, and a "work based on the Program" means either the Program or any derivative work under copyright law: that is to say, a work containing the Program or a portion of it, either verbatim or with modifications and/or translated into another language. (Hereinafter, translation is included without limitation in the term "modification.") Each licensee is addressed as "you."

Activities other than copying, distribution and modification are not covered by this License; they are outside its scope. The act of running the Program is not restricted, and the output from the Program is covered only if its contents constitute a work based on the Program (independent of having been made by running the Program). Whether that is true depends on what the Program does.

1. You may copy and distribute verbatim copies of the Program source code as you receive it, in any medium, provided that you conspicuously and appropriately publish on each copy an appropriate copyright notice and disclaimer of warranty; keep intact all the notices that refer to this License and to the absence of any warranty; and give any other recipients of the Program a copy of this License along with the Program.

You may charge a fee for the physical act of transferring a copy, and you may at your option offer warranty protection in exchange for a fee.

2. You may modify your copy or copies of the Program or any portion of it, thus forming a work based on the Program, and copy and distribute such modifications or work under the terms of Section 1 above, provided that you also meet all of these conditions:

> a) You must cause the modified files to carry prominent notices stating that you changed the files and the date of any change.

> b) You must cause any work that you distribute or publish, that in whole or in part contains or is derived from the Program or any part thereof, to be licensed as a whole at no charge to all third parties under the terms of this License.

> c) If the modified program normally reads commands interactively when run, you must cause it, when started running for such interactive use in the most ordinary way, to print or display an announcement including an appropriate copyright notice and a notice that there is no warranty (or else, saying that you provide a warranty) and that users may redistribute the program under these conditions, and telling the user how to view a copy of this License. (Exception: if the Program itself is interactive but does not normally print such an announcement, your work based on the Program is not required to print an announcement.)

These requirements apply to the modified work as a whole. If identifiable sections of that work are not derived from the Program, and can be reasonably considered independent and separate works in themselves, then this License, and its terms, do not apply to those sections when you distribute them as separate works. But when you distribute the same sections as part of a whole which is a work based on the Program, the distribution of the whole must be on the terms of this License, whose permissions for other licensees extend to the entire whole, and thus to each and every part regardless of who wrote it.

Thus, it is not the intent of this section to claim rights or contest your rights to work written entirely by you; rather, the intent is to exercise the right to control the distribution of derivative or collective works based on the Program.

In addition, mere aggregation of another work not based on the Program with the Program (or with a work based on the Program) on a volume of a storage or distribution medium does not bring the other work under the scope of this License.

3. You may copy and distribute the Program (or a work based on it, under Section 2) in object code or executable form under the terms of Sections 1 and 2 above provided that you also do one of the following:

> a) Accompany it with the complete corresponding machine-readable source code, which must be distributed under the terms of Sections 1 and 2 above on a medium customarily used for software interchange; or,
>
> b) Accompany it with a written offer, valid for at least three years, to give any third party, for a charge no more than your cost of physically performing source distribution, a complete machine-readable copy of the corresponding source code, to be distributed under the terms of Sections 1 and 2 above on a medium customarily used for software interchange; or,
>
> c) Accompany it with the information you received as to the offer to distribute corresponding source code. (This alternative is allowed only for noncommercial distribution and only if you received the program in object code or executable form with such an offer, in accord with Subsection b above.)

The source code for a work means the preferred form of the work for making modifications to it. For an executable work, complete source code means all the source code for all modules it contains, plus any associated interface definition files, plus the scripts used to control compilation and installation of the executable. However, as a special exception, the source code distributed need not include anything that is normally distributed (in either source or binary form) with the major components (compiler, kernel, and so on) of the operating system on which the executable runs, unless that component itself accompanies the executable.

If distribution of executable or object code is made by offering access to copy from a designated place, then offering equivalent access to copy the source code from the same place counts as distribution of the source code, even though third parties are not compelled to copy the source along with the object code.

4. You may not copy, modify, sublicense, or distribute the Program except as expressly provided under this License. Any attempt otherwise to copy, modify, sublicense or distribute the Program is void, and will automatically terminate your rights under this License. However, parties who have received copies, or rights, from you under this License will not have their licenses terminated so long as such parties remain in full compliance.

5. You are not required to accept this License, since you have not signed it. However, nothing else grants you permission to modify or distribute the Program or its derivative works. These actions are prohibited by law if you do not accept this License. Therefore, by modifying or distributing the Program (or any work based on the Program), you indicate your acceptance of this License to do so, and all its terms and conditions for copying, distributing or modifying the Program or works based on it.

6. Each time you redistribute the Program (or any work based on the Program), the recipient automatically receives a license from the original licensor to copy, distribute or modify the Program subject to these terms and conditions. You may not impose any further restrictions on the recipients' exercise of the rights granted herein. You are not responsible for enforcing compliance by third parties to this License.

7. If, as a consequence of a court judgment or allegation of patent infringement or for any other reason (not limited to patent issues), conditions are imposed on you (whether by court order, agreement or otherwise) that contradict the conditions of this License, they do not excuse you from the conditions of this License. If you cannot distribute so as to satisfy simultaneously your obligations under this License and any other pertinent obligations, then, as a consequence, you may not distribute the Program at all. For example, if a patent license would not permit royalty-free redistribution of the Program by all those who receive copies directly or indirectly through you, then the only way you could satisfy both it and this License would be to refrain entirely from distribution of the Program.

If any portion of this section is held invalid or unenforceable under any particular circumstance, the balance of the section is intended to apply and the section as a whole is intended to apply in other circumstances.

It is not the purpose of this section to induce you to infringe any patents or other property right claims or to contest validity of any such claims; this section has the sole purpose of protecting the integrity of the free software distribution system, which is implemented by public license practices. Many people have made generous contributions to the wide range of software distributed through that system in reliance on consistent application of that system; it is up to the author/donor to decide if he or she is willing to distribute software through any other system and a licensee cannot impose that choice.

This section is intended to make thoroughly clear what is believed to be a consequence of the rest of this License.

8. If the distribution and/or use of the Program is restricted in certain countries either by patents or by copyrighted interfaces, the original copyright holder who places the Program under this License may add an explicit geographical distribution limitation excluding those countries, so that distribution is permitted only in or among countries not thus excluded. In such case, this License incorporates the limitation as if written in the body of this License.

9. The Free Software Foundation may publish revised and/or new versions of the General Public License from time to time. Such new versions will be similar in spirit to the present version, but may differ in detail to address new problems or concerns.

Each version is given a distinguishing version number. If the Program specifies a version number of this License which applies to it and "any later version," you have the option of following the terms and conditions either of that version or of any later version published by the Free Software Foundation. If the Program does not specify a version number of this License, you may choose any version ever published by the Free Software Foundation.

10. If you wish to incorporate parts of the Program into other free programs whose distribution conditions are different, write to the author to ask for permission. For software which is copyrighted by the Free Software Foundation, write to the Free Software Foundation; we sometimes make exceptions for this. Our decision will be guided by the two goals of preserving the free status of all derivatives of our free software and of promoting the sharing and reuse of software generally.

NO WARRANTY

11. BECAUSE THE PROGRAM IS LICENSED FREE OF CHARGE, THERE IS NO WARRANTY FOR THE PROGRAM, TO THE EXTENT PERMITTED BY APPLICABLE LAW. EXCEPT WHEN OTHERWISE STATED IN WRITING THE COPYRIGHT HOLDERS AND/OR OTHER PARTIES PROVIDE THE PROGRAM "AS IS" WITHOUT WARRANTY OF ANY KIND, EITHER EXPRESSED OR IMPLIED, INCLUDING, BUT NOT LIMITED TO, THE IMPLIED WARRANTIES OF MERCHANTABILITY AND FITNESS FOR A PARTICULAR PURPOSE. THE ENTIRE RISK AS TO THE QUALITY AND PERFORMANCE OF THE PROGRAM IS WITH YOU. SHOULD THE PROGRAM PROVE DEFECTIVE, YOU ASSUME THE COST OF ALL NECESSARY SERVICING, REPAIR OR CORRECTION.

12. IN NO EVENT UNLESS REQUIRED BY APPLICABLE LAW OR AGREED TO IN WRITING WILL ANY COPYRIGHT HOLDER, OR ANY OTHER PARTY WHO MAY MODIFY AND/OR REDISTRIBUTE THE PROGRAM AS PERMITTED ABOVE, BE LIABLE TO YOU FOR DAMAGES, INCLUDING ANY GENERAL, SPECIAL, INCIDENTAL OR CONSEQUENTIAL DAMAGES ARISING OUT OF THE USE OR INABILITY TO USE THE PROGRAM (INCLUDING BUT NOT LIMITED TO LOSS OF DATA OR DATA BEING RENDERED INACCURATE OR LOSSES SUSTAINED BY YOU OR THIRD PARTIES OR A FAILURE OF THE PROGRAM TO OPERATE WITH ANY OTHER PROGRAMS), EVEN IF SUCH HOLDER OR OTHER PARTY HAS BEEN ADVISED OF THE POSSIBILITY OF SUCH DAMAGES.

END OF TERMS AND CONDITIONS

Appendix: How to Apply These Terms to Your New Programs

If you develop a new program, and you want it to be of the greatest possible use to the public, the best way to achieve this is to make it free software which everyone can redistribute and change under these terms.

To do so, attach the following notices to the program. It is safest to attach them to the start of each source file to most effectively convey the exclusion of warranty; and each file should have at least the "copyright" line and a pointer to where the full notice is found.

> one line to give the program's name and a brief idea of what it does.
> Copyright © 19yy name of author
>
> This program is free software; you can redistribute it and/or modify it under the terms of the GNU General Public License as published by the Free Software Foundation; either version 2 of the License, or (at your option) any later version.
>
> This program is distributed in the hope that it will be useful, but WITHOUT ANY WARRANTY; without even the implied warranty of MERCHANTABILITY or FITNESS FOR A PARTICULAR PURPOSE. See the GNU General Public License for more details.
>
> You should have received a copy of the GNU General Public License along with this program; if not, write to the Free Software Foundation, Inc., 675 Mass Ave, Cambridge, MA 02139, USA.

Also add information on how to contact you by electronic and paper mail.

If the program is interactive, make it output a short notice like this when it starts in an interactive mode:

> Gnomovision version 69, Copyright © 19yy name of author
> Gnomovision comes with ABSOLUTELY NO WARRANTY; for details type 'show w'.
>
> This is free software, and you are welcome to redistribute it under certain conditions; type 'show c' for details.

The hypothetical commands 'show w' and 'show c' should show the appropriate parts of the General Public License. Of course, the commands you use may be called something other than 'show w' and 'show c'; they could even be mouse-clicks or menu items—whatever suits your program.

You should also get your employer (if you work as a programmer) or your school, if any, to sign a "copyright disclaimer" for the program, if necessary. Here is a sample; alter the names:

> Yoyodyne, Inc., hereby disclaims all copyright interest in the program 'Gnomovision' (which makes passes at compilers) written by James Hacker.
>
> signature of Ty Coon, 1 April 1989
> Ty Coon, President of Vice

This General Public License does not permit incorporating your program into proprietary programs. If your program is a subroutine library, you may consider it more useful to permit linking proprietary applications with the library. If this is what you want to do, use the GNU Library General Public License instead of this License.

MODIFIED BSD LICENSE

Redistribution and use in source and binary forms, with or without modification, are permitted provided that the following conditions are met:

1. Redistributions of source code must retain the above copyright notice, this list of conditions and the following disclaimer.
2. Redistributions in binary form must reproduce the above copyright notice, this list of conditions and the following disclaimer in the documentation and/or other materials provided with the distribution.
3. The names of the authors may not be used to endorse or promote products derived from this software without specific prior written permission.

THIS SOFTWARE IS PROVIDED ``AS IS" AND WITHOUT ANY EXPRESS OR IMPLIED WARRANTIES, INCLUDING, WITHOUT LIMITATION, THE IMPLIED WARRANTIES OF MERCHANTABILITY AND FITNESS FOR A PARTICULAR PURPOSE.

Index

A
access
  ACL approach to controlling, 61–62
  blocking to MSRating.dll, 98–99
  control models generally, 59–60
  programmatic, to applications, 92
  unauthorized, 23–24
Access server, system table names (table), 314
ACL (access control list) approach to access control, 61–62
Active Server Pages, safe HTML encoding, 275
ActiveX controls
  and cross-site scripting vulnerabilities, 270, 272
  safe for scripting, 91
  unsafe, 328–329
AdCycle, 324
Address Resolution Protocol. *See* ARP
Aitel, Dave, 47, 138
alert () function (JavaScript), 271
Alexis, 199
anti-bugging, applications that implement, 97
API monitors, 49–50
Apple QuickTime, 183
application heap overflows, 116–122
application layer
  in multilayer model, 33–34
  spoofing at, 250–251
  vulnerabilities in, 335
applications
  attackers taking control of, 108
  password attack protection, 83–84
  permissions set on, 204, 207
  programmatic access to, 88, 92
  subject to JavaScript manipulation, 91
  that implement anti-debugging, 97
  Web, testing, 47
AppScan security scanner, 53
arithmetic operations on pointers, 178
ARP (Address Resolution Protocol) cache poisoning, 29–30
ASP pages, safe HTML encoding, 275
attackers vs. hackers vs. crackers, 20–23
attacks
  buffer overflow. *See* buffer overflow vulnerabilities
  C functions vulnerable to format string (table), 157
  denial-of-service. *See* DoS attacks
  dynamic linking and loading, 95–100
  man-in-the-middle. *See* man-in-the-middle attacks
  phishing, 20, 22, 88, 249–250
  resources about, 23
  session replay, 280
  spoofing. *See* spoofing
  teardrop, 46
  topographical, 330
  and user input vulnerabilities, 7
authentication, Web server vs. user, 295
Authenticode signing capability, 101
authors of this book, 17–18

B
back doors
  described, 22
  in URLs, 330
BackOriface, 20
Bad Software (Kaner and Pels), 25
banners, vulnerability testing, 39
Baseline Security Analyzer, 39–40
Bell-Lapadula access control model, 60–62
Bernstein, Daniel J., 10
binary formats, vulnerabilities of, 137–138
binary scanners, 48–49
bitmap format, 135–136
black-box testing for buffer overflow vulnerabilities, 129–130
black hat attackers
  described, 23
  and security testers, 37–38
Black, Uyless, 35
block-based fuzzing, 138
book, this
  authors, 17–18
  CD-ROM. *See* CD-ROM
  message of, 10–13
  structure of, 13–17
Bouchareine, Pascal, 157
browser-based phishing attacks, 88
Browser Helper Objects, 331
browser-hosted applications, 91
browsing, forceful, 277–283
buffer overflow vulnerabilities
  exploiting, 122–127
  finding and fixing, 127–130
  generally, 107–109
  heap overflows, 116–122
  stack overflows, 109–116
buffer overruns, 67–68
BugScan, 49
BugTraq mailing list, 338
building
  dictionary helper, 78–80
  password cracker, 75–83
  proxy-based fuzzer, 139
Building Secure Software (McGraw), 17

C
C, C++
  and buffer overflow attacks, 108–109
  functions vulnerable to format string attack (table), 157
  integer overflow vulnerabilities, 173–179
  source scanning tools, 128
  unsafe routines and alternatives (table), 130
cache poisoning
  ARP, 29–30
  DNS, 250
  name server, 247–249
California's identify theft prevention law, 302
CAPS lock, 261
Carroll, Lewis, 221
CD-ROM
  example source code on, 13
  HexDump application, 109
  Memsrch.cpp, 215
  password generator, 84
  programs contained, installing, 341–348
  SimpleFuzz, 139
CERT (Computer Emergency Response Team)
  number of recent security incidents (fig.), 336
  online vulnerability database, 12
  vulnerabilities reported to in recent years, 5–6
  Web site, 338
certification, security, 339
Certified Information Systems Security Professional (CISSP), 339
CGI (Common Gateway Interface) programs, 196, 251, 269, 278, 280
Chase, Scott, 18
checksum of a DLL, 101
Cigital's ITS4 security scanner, 128
CIL file format vulnerabilities, 137–138
CISSP certification, 339
client-side input validation
  circumventing, 273
  and SQL injections, 314
clipart, vulnerable formats, 137–138
code on CD-ROM, 13
Code Red worm, 127, 132
ColdFusion, 278, 281
command interpreters, embedded script languages and, 89
commands, common SQL (table), 312
Common Gateway Interface (CGI), 251, 269, 278, 280
Common Vulnerabilities and Exposures (CVE) database, 157
computer security
  hackers vs. crackers vs. attackers, 20–23
  legal and ethical issues, 23–26
  sources of information, 338–339

349

350 Index

confidentiality
　privacy in Bell-Lapadula access control model, 61
　and swap file vulnerabilities, 223–226
　using disk editor to find confidential fragments, 226–230
connections
　hijacking, 240–247
　proxying between, 140–141
Content Advisor, Internet Explorer, 97–98
cookies
　and information protection, 301–302
　poisoning, 270–272
　using on Web sites, 308
CPU resources
　swap files and, 223–224
　Unix timesharing, 62–64
crack () function, 80–83
crackers
　password, 75–83
　vs. hackers vs. attackers, 20–23
cracking tools, 44–47
crashed machines, testing for, 42
credentials, DRM's use of, 214
cross-site scripting
　finding vulnerabilities, 271–273
　fixing vulnerability, 274–275
　vulnerability described, 267–270
crypt () function, 45, 197
CryptCreateHash, CryptHashData functions, 198
cryptography
　and hash functions, 195
　Windows Cryptographic API, 198
customer information, protecting, 302
CVE database, 157

D

damage, federal legal definition, 24
data link layer in multilayer model, 28
data types, names and ranges of common C, C++ (table), 174
database servers, system table names (table), 314
databases, stored procedures in, 91
DataRescue, Inc., 50
de-escalation bugs, 62
debuggers, using for security testing, 50–52
default passwords. See weak passwords
DEF.CON event, 339
DeleteFile () function, 226
denial-of-service attacks. See DoS attacks
DES encryption, 197
developers' need to address vulnerabilities, 5–7
DHCP (Dynamic Host Configuration Protocol) and man-in-the-middle attacks, 251–252
dictionary helper, building, 78–80
Digital Millennium Copyright Act (DMCA), 24
Digital Rights Management (DRM), 97, 100, 202, 213–214
disassemblers, 50
Disk Doctor (Norton), 226
disk fragmentation, finding confidential fragments, 226–230
disk system vulnerabilities, 225–226
DLLs (dynamic link libraries), explicit linking and protecting against attacks, 101–102

DNS cache poisoning, 250
DNS spoofing, 249–250
Document Object Model, 270, 271
documents, macro commands embedded within, 90
DoHexDump function, 110
Domain Name Server (DNS) protocol, 32
domain names, spoofing, 249–250
DoS (denial-of-service) attacks
　and EPS, 137
　and format string vulnerabilities, 159
　teardrop attacks, 46
download sites
　BugScan, 49
　BugTraq mailing list, 338
　Ethereal sniffer program, 43
　Holodeck API analysis tool, 98, 100
　LC4 password cracker, 75
　MD4 hashing algorithm, 81
　Spike network fuzzer, 138
Dr. Dobbs Journal, 339
DRM (Digital Rights Management), 97, 100, 202, 213–214
dtappgather vulnerability, 67
Dynamic Host Configuration Protocol (DHCP) and man-in-the-middle attacks, 251–252
dynamic linking and loading
　finding vulnerability, 100–101
　fixing vulnerability, 101–103
　vulnerability of, 95–100

E

e-mail
　and phishing, 249–250
　temporary files vulnerability, 206
Economic Espionage Act, 24
editors
　disk, using to find confidential fragments, 226–230
　regedit registry, 191
　specialty, 49
eEye Digital Security Team, 53
embedded script languages and command interpreters, 89
Encapsulated PostScript (EPS) format, 135–136
encapsulation, TCP/IP, typical (fig.), 27
encryption
　passwords, 187
　in Windows, 232
EPS (Encapsulated PostScript) format, 135–136
error handlers, 261
error messages
　common HTML codes, 289
　overly revealing, 255–256
escalation of privilege
　bugs, 62
　described, 21–22
　vulnerabilities, fixing on Windows, 68–71
escape characters in Web applications, 324
Ethereal, installing, 343
Ethereal sniffer program, 43–44
EtherPeek, 38
ethical hackers described, 22
ethical issues surrounding computer security, 23–26
Ettercap, 38, 43, 248–249, 253, 343

execute permissions in Unix permissions scheme, 63
explicit linking and loading DLLs, 101–102
exploiting
　format string vulnerabilities, 158–168
　integer overflow vulnerabilities, 179
　Web sites through SQL injection, 316–319
exploits
　See also specific exploit
　buffer overflow, 122–127
　described, 21
　stack overflows, 113–116
Extended Stack Pointer (ESP) register and stack overflow exploit, 113

F

federal laws related to illegal computer use, 23–26
File Allocation Table (FAT), ghosts of deleted files (fig.), 229
file formats
　finding vulnerabilities with fuzzing, 138–147
　vulnerabilities of, 133–138
file sharing utility Samba, 258
Filemon (Sysinternal), 204
files
　Goggling for hidden, 283
　swap or paging, 223
　temporary, vulnerability of, 201–205
filesystem interface, checking for long string inputs, 129
Firefox browser spoofing attempt (fig.), 247
firewalls described, 22, 28
FishCart, 183
fixing
　See also preventing
　buffer overflow vulnerabilities, 130
　cross-site scripting vulnerability, 274–275
　dynamic linking and loading vulnerabilities, 101–102
　format string vulnerabilities, 168–170
　integer overflow vulnerabilities, 181
　leaving things in memory, 221
　password vulnerabilities, 83–84
　preventing spoofing and man-in-the-middle attacks, 252
　proprietary formats, protocols vulnerabilities, 147–148
　script and macro vulnerabilities, 92–93
　SQL injection vulnerabilities, 322–323
　supervisor bit set vulnerability, 69–70
　swap files and incomplete deletes, 230–232
　temporary files vulnerability, 207
Flawfinder GPL vulnerability finder, 128
forceful browsing
　building test tool, 283–294
　preventing, 295
　vulnerability described, 277–283
form data manipulation
　described, 302–306
　preventing, 307–309
format functions and format strings, 151–153
format specifiers, 153–154
format strings
　in C, vulnerable to attack (table), 157
　vulnerabilities, 151–156

Index

formats
    IP packet (fig.), 33
    proprietary, vulnerability of, 133–138
    TCP packet (fig.), 33
    UDP packet (fig.), 34
forms, Web. *See* Web forms
401 Authorization Required error messages, 258–260
fprintf function, 156
fragmentation, disk, 226
fread () function, 225
full disclosure approach described, 26
functions
    See also specific function
    C, C++ prone to buffer overflows (table), 109
    format, 151
    hash. *See* hashing function, hash function
FuzzBrowse, 307
fuzzing to find vulnerabilities, 138–147
fuzzing, network, 46–47

G

gateways subnetting and, 31
GDB debugger, 52
generating memorable passwords, 84–85
GerProcAddress functions, 102
getpass () function, 213, 221
gets function, 118, 121
GNU Anubis, 171
GNU General Public License, 349–354
Goggling for hidden files, 283
Guniski, Georgi, 332

H

hackers
    romance and reality of, 19–20
    vs. crackers vs. attackers, 20–23
hacking tools, 44–47
hash functions
    and user authentication, 188
    using, 195–196
hashing function, Windows, 81
HBGary, 49
heaps
    overflows, 116–122
    typical structure of (fig.), 117
'Hello' bug, 138
HexDump application, 109
hidden form fields, 309
hijacking connections, 240–247, 253
Holodeck API analysis tool, 53–54, 98, 204
hosts on networks, identifying, 38
How to Break Software Security (Thompson & Whittaker), 10
Howard, Michael, 6
HTML code, and cross-site scripting, 267–271
HTML-encoded output, 275
HTML error codes, 289
HTML Help, 91
HTTP (HyperText Transfer Protocol)
    data passed to server, 299
    protocol described, 35
hubs described, 27
Huseby, Sverre, 17

I

'I Love You' virus, 88, 92, 94

IANA Port List, 134
ICMP (Internet Control Message Protocol)
    echo request messages, ping and, 42
    protocol described, 35
identify theft, 302
identifying hosts on networks, 38
IGMP (Internet Group Management Protocol) described, 35
imaemap.exe, 262
image formats, 135–136
information disclosure
    customer, and identify theft, 302
    described, 21
    vulnerability, finding and fixing, 255–263
initializing variables, 211
injection attacks
    analyzing APIs for, 53–54
    SQL, 47, 256, 311–324
Innocent Code (Huseby), 17
Inoculan AV client, 199
input
    validating, 7–8
    circumventing validation, , 273
installing programs on CD-ROM, 341–348
integer overflow vulnerabilities
    described, 173–179
    exploiting, finding, fixing, 179–181
integer underflows, 179
interfaces
    building for forceful browsing test tool, 283–284
    checking for long string inputs, 129
Internet Control Message Protocol. *See* ICMP
Internet Explorer
    browser security issues, 331–332
    Content Advisor, 97
    cross-site scripting vulnerability, 276
    listing libraries in, 98
    man-in-the-middle attack vulnerability, 253
    and phishing, 250
Internet Group Management Protocol (IGMP), 35
Internet Information Server (IIS)
    and Code Red worm, 127
    scripting vulnerability, 272
Internet Protocol. *See* IP
intrusion detection systems (IDSs), 22, 28
IP (Internet Protocol) described, 30
IP packets, format of (fig.), 30
IP spoofing, 247
ITS4 security scanner, 128

J

J2EE, 278
JavaScript
    and cross-site scripting, 267
    manipulation of, 91
    <script> tag, 271
    and URL spoofing, 329
John the Ripper password cracker, 45, 196, 344
Johnson, Angus, 49
Jpcap, 43
JPEG format, 135–136

K

Kaner, Cem, 25
keystroke logging, 96

kill command in Apache servers, 70
Knoppix, 227

L

L0phtcrack tool, 45
LC4 password cracker, 75–76
LeBlanc, David, 6
leetspeek described, 21
legal issues surrounding computer security, 23–26
Libnet, installing, 342–343
libnet tool, 46
Libpcap library, Ethereal and Tcpdump, 43
library search order and dynamic linking attack, 96
license
    GNU General Public, 349–354
    software agreements prohibiting reverse engineering, 25
limits.h, 174
linking, explicit, 101
Linux
    GDB debugger, 52
    memory locking, 231
    searching memory with custom debugger, 220–221
    seteuid () function, 70
    shadow file (fig.), 77
LoadLibrary function, 102
LoadLibrary systems, LoadLiba, 98
logical names, mapping, 32
LogicLibrary, 49
logs, macro expansion in, 91–92
long string insertions corruptor, 139, 145, 181

M

MAC addresses
    described, 29
    spoofing, 252
    and switches, 28
macro expansion in logs, messages, 91–92
Macromedia's ColdFusion, 278, 281
macros and system-level attacks, 87–88
man-in-the-middle attacks
    described, 237–238
    DHCP and 802.11, 251–252
    finding, 238–252
masquerading described, 28, 30
MAXLENGTH field, 273
McGraw, Gary, 17
MD4, MD5 hashing algorithms, 80–83, 195, 196
Media Access Control addresses. *See* MAC addresses
Melissa virus, 88, 94
memory
    and buffer overflow attacks, 108
    leaving things in, vulnerability, 211–221
    Memsrch.cpp search tool, 215
    maps, 224
    protection, and finding exposed memory, 214–215
memory management unit (MMU), 224
memset () function, 221
Memsrch.cpp, 215
Message Digest algorithm. *See* MD4, MD5 hashing algorithm
messages, macro expansion in, 91–92
Metasploit, 107–108

Microsoft Managed Code, 16
Microsoft Baseline Security Analyzer, 39–40
Microsoft Clipart Gallery file format vulnerability, 137–138
Microsoft Help and Support Center, 331
Microsoft MSHTML object, 283
Microsoft Notebook, 113
Microsoft Office, 94
Microsoft Query Analyzer, 138
Microsoft SQL Server
    CAN-2002-1123 vulnerability, 149
    passwords on, 75
    vulnerabilities of, 138
MIME file type, 35
MIME-conversion buffer overflow, 132
MIME-type spoofing, 330
models
    Bell-Lapadula access control, 60–62
    dynamic linking and loading, 95–96
    multilayer, described, 27–35
    network, generally, 59–60
monitors, API and system, 49–50
Morris Internet worm, 86
MS00-15 vulnerability, 149
MSRating.dll, 98
multilayer model
    layers described, 28–35
    and TCP/IP, 27
multiprivilege system issues, 62
music files, playback, 202
MySQL database engine, 188, 268

N
name server cache poisoning, 247–249
Nemesis tool, 46, 345
Nessus security scanner, 39–40, 345
Netcat program, 45–46
Network Address Translation (NAT) and masquerading, 28
network fuzzing, 46–47
network interface, checking for long string inputs, 129
network layer in multilayer model, 29–30
network models generally, 59–60
network scanner, Nmap, 41–42
network security paradigm driving market, 3
networking, basics of, 26–35
networks, identifying hosts on, 38
nextWord () function, pwDictionary class, 80
Nikto Web site test tool, 47, 345
Nmap
    installing, 346
    and network scanners, 41–42
NOP (no operation) sled technique, 165–166
Norton's Disk Doctor, 226
Notepad and stack overflow exploit, 113
NSA's Center of Excellence program, 10
NT Symbolic Debugger (NTSD), 51–52
ntsd debugger, 114–115
Nullsoft Winamp media player, 188
NUM lock, 261

O
objects
    in access control models, 60
    Browser Helper Objects, 331
ODBC error messages and SQL injection attacks, 256–257, 316
OllyDbg debugger, 51, 85, 160–161, 191–195
online vulnerability database, SANS, 47
open relay e-mail servers, 250–251
Open Web Application Security Project (OWASP) bug list, 47
OpenSSL Blowfish algorithm, 224–225
Oracle system table names (table), 314
oracles, and security testing, 43–44
Ornaghi, Alberto, 43
OS command injection attacks, 47
OS fingerprinting described, 42
output, HTML-encoded, 275
OWASP (Open Web Application Security Project) 'top ten' bugs list, 47
ownership of system resources, 62

P
Pablo Software Solutions, 258
packet
    generation and replay, 45–46
    sniffing tools, 42–44
packets
    IP. See IP packets
    transmission of, 30–31
pagefile.sys, 227
paging files, 223
parameters
    identifying forceful browsing vulnerabilities in, 282
    parsing for file access, 295
password crackers, 75–83
passwords
    See also permissions
    and cracking, 22
    cracking tools, 45
    generating memorable, 84–85
    inside applications, 212–213
    left in memory, 214
    phishing for, 249–250
    storing in plaintext, vulnerability, 187–196
    strong, enforcing use, 83
    using hash functions, 195–196
    weak, vulnerabilities of, 73–75, 83
PE file editors, 49
Pels, David, 25
penetration testing, 22
'%Login invalid' message, 264
Perl code search request, 274
permissions
    See also passwords
    access control models, 59–60
    misconfiguration issues, 62
    set on applications, 204
    on temporary files, 207
Pescatore, Joe, 17
phishing attacks, 20, 22, 88, 249–250
PHP magic quotes, 324
physical layer in multilayer model, 28
physical security described, 16
pin numbers, Web forms' use of, 297–301
ping program, using to test for crashed machine, 42
plug-ins
    See also specific plug-in
    described, 39–40
PNG (Portable Network Graphics) files, 147–148
pointers, arithmetic operations on, 178
port 80, assignment of, 33
port numbers
    common TCP application (table), 34
    IANA Port List, 134
port scanners described, 41–42
Portable Network Graphics file. See PNG files
PostScript, Encapsulated format, 135–136
preventing
    See also fixing
    cross-site scripting vulnerabilities, 275
    forceful browsing, 295
    spoofing and man-in-the-middle attacks, 252
printf () function
    format function, 151–153
    and format string vulnerabilities, 158, 169–170
    and static linking, 95
privacy. See confidentiality
privileges
    elevated, Summary Sheet, 70–71
    escalation of, 21–22
    multiprivilege system issues, 62
    root, and setuid () function, 69–70
product security, five things to improve, 11–12
products and default passwords, 75, 83
programmatic
    access, applications supporting, 92
    interfaces, checking for long string inputs, 129
programming languages
    high-level, and memory leaks, 212
    Nessus Attack Scripting Language (NASL), 39–40
    script language vulnerabilities, 89
    scripting implementation in, 93
    and vulnerabilities, 13
programs
    on CD-ROM, installing, 341–348
    supervisor-enabled, 64–68
promiscuous mode, sniffers, 42–43
proprietary formats, preventing problems with, 147–148
protocols
    See also specific protocol
    finding vulnerabilities with fuzzing, 138–147
    proprietary, preventing problems with, 147–148
    TCP application, and port numbers (table), 34
proxy-based fuzzing, 139
ptrace () function, 220
pwdump3 utility, 77

Q
Query Analyzer, 138
query strings, vulnerability, 306
Quick 'n Easy FTP Server 1.77, 258
QuickTime, Apple, 183

R
RATS (rough auditing tool for security)
    installing, 346–347
    using, 48–49, 128
RDISK utility, 202–205
read permissions in Unix permissions scheme, 63
ReadProcessMemory function, 219
red teaming described, 22

regedit registry editor, 191
registry, editing to delete page files, 231
Regmon API monitoring tool, 188–191
remote procedure calls (RPCs) and UDP, 33
remove () function, 226
replay, packet generation and, 45–46
replay attacks
　finding, 294
　session, 280
reporting vulnerabilities, 26
requests, HTTP
　described, 35
　and field manipulation, 299–301
Resource Hacker tool, 49
Retina security scanner, 53
reverse engineering
　described, 22
　for security testing, 24–25
　tools, 47–52
root shell and buffer overruns, 67–68
rough auditing tool for security. *See* RATS
routers described, 28
routines, unsafe C (table), 130

S

'safe for scripting' ActiveX controls, 91
SAINT, 38
salt, and hashed passwords, 196
Samba SWAT interface, 257
'sandbox,' preventing malicious behavior with, 93
SANS Web application bug list, 47
SATAN (Security Administrator Tool for Analyzing Networks)
　installing, 347
　scanner described, 39
scanners
　binary, source, 48–49
　security, 38–44
script kiddies described, 21
<script> tag (JavaScript), 271
scripts
　cross-site. *See* cross-site scripting
　and system-level attacks, 87–88
　vulnerabilities, 88
　writing safe, 92–93
Sealy, Don, 86
search
　Goggling for hidden files, 283
　library search order, and dynamic linking attack, 96
　pages, and cross-site scripting, 268–270
searching
　Linux memory with custom debugger, 220–221
　Windows memory with custom debugger, 215–220
Secure Hash Algorithm (SHA-1), 195
SecureZeroMemory () function, 221
security
　call to action for developers, 3–10
　and complex formats, 135
　contextual nature of, 9
　incident reported (fig.), 336
　product. *See* product security
　sources of more information, 338–339
Security Administrator Tool for Analyzing Networks (SATAN) scanner, 39, 347
Security Engineering (Anderson), 16
Security IDs (SIDs) and escalation of privilege, 69–70

Security Innovation, Holodeck API analysis tool, 100, 204
security scanners types, using, 38–44
security testers
　described, 22–23
　use of reverse engineering, 24–25
security testing
　commercial tools, 53–54
　hacking, cracking tools, 44–47
　reverse engineering tools, 47–52
　security scanners, 38–44
　tools generally, 37–38
SecurityFocus, 12, 233
Server Side Includes (SSIs) within Web pages, 90
session cookies, 270–272
session replay attacks, 280
set_key () function, 225
setFile () function, pwDictionary class, 80
setuid () functions family, 63, 69–70
SHA-1 (Secure Hash Algorithm), 195
shell code
　anatomy of malicious, 5
　Windows message box created by (fig.), 4
shells
　and system-level attacks, 87–88
　and uncommon URL schemes, 330–331
side-effect functionality
　and buffer overruns, 67–68
　exploiting programs, 64–67
signed integers, 148, 176
Simple Mail Transport Protocol. *See* SMTP
SimpleFuzz, 139–140
SMTP (Simple Mail Transport Protocol)
　described, 33
　and spoofing, 250
　vulnerability of, 148
sniffers described, 42–43
social engineering
　described, 16–17, 22
　phishing, 249–250
SoftICE debugger, 52
software
　See also specific programs
　license, GNU, 349–354
　licenses prohibiting reverse engineering, 25
　need for secure, 5–7
source code
　on CD-ROM, 13
　and reverse engineering, 47
　scanners, 48–49, 128
spammers, 250
specialty editors, 49
Spike network fuzzer, 47, 138
spoofing
　ARP, 30
　attacks described, 237–240
　connection hijacking, 240–247
　described, 21, 237–240
　name server cache poisoning, 247–249
　URLs, 329–331
sprintf function, 50, 156
SQL commands, special characters (table), 312
SQL injection attacks, 47, 256, 311–324
SQLSnake vulnerability, 75, 86
stack, memory, and buffer overflow attacks, 109–113
stack overflows
　described, 109–113

exploiting, 113–116
static analyzers, 48–49
static linking described, 95
Staub, Phil, 77
Stevens, W. Richard, 27, 35
storing passwords
　in plaintext, 187–196
　using hash functions, 195
strace function, 50
strcpy function, 50, 118, 128
streaming media and Nullsoft Winamp media player, 188
strings
　long insertions, 139
　query, 306
strong passwords, enforcing, 83
subnetting described, 31
substitution corruption
　described, 139
　in SimpleFuzz, 146–147
suid bugs, 62, 69–70
Summary Sheets
　additional browser security issues, 331–332
　buffer overflows, 131
　cross-site scripting, 276
　dynamic linking and loading, 103
　forceful browsing, 295–296
　format for, 14
　format string vulnerabilities, 170–171
　integer overflows, 182–183
　leaving things in memory, 221–222
　parameter tampering, cookie poisoning, hidden field tampering, 309–310
　permitting weak passwords, 85–86
　proprietary formats, protocols, 148–149
　running with elevated privilege, 70–71
　shells, scripts, macros, 93–94
　spoofing and man-in-the-middle attacks, 252–253
　SQL injection vulnerabilities, 323–324
　storing passwords in plain text, 198
　swap files and incomplete deletes, 232–233
　temporary files vulnerability, 207–209
　volunteering excess information, 263–264
superuser access control model, 61–62
supervisor bit set
　avoiding, 69–70
　discovering if program has, 64
supervisor-enabled programs, 64–68
swap files and incomplete deletes
　fixing vulnerability, 230–232
　vulnerability described, 223–230
switches described, 28
Sysinternal's Filemon utility, 204
Sysinternals.com, 50
syslog function, 156
syslog-ng product, 92
system-level attacks
　permissions problems, 59–60
　shells, scripts, macros, 87–88
　weak passwords, 73–75
system monitors, 49–50
system requirements for CD-ROM, 342
systems in access control models, 60

T

Tabular Data Stream format (TDS), 138
TCP application port numbers (table), 34

TCP/IP protocol described, 26–27
TCP/IP and Related Protocols (Black), 35
TCP/IP Illustrated (Stevens), 35
TCP port 80, 41
Tcpdump packet sniffer, 43, 46, 347–348
tcpreplay, tcpslice tools, 46
teardrop attacks, 46
temporary files vulnerability
    described, 201–205
    finding, fixing, 206–207
testing
    for buffer overflow vulnerabilities, 127–129
    for crashed machine, 42
    for forceful browsing, 283–294
    for integer overflows, 180
    passwords against password database, 77–78
text formats, vulnerabilities of, 137–138
Thompson, Herbert, 17–18
three-way handshake described, 32
Through the Looking Glass (Lewis), 221
'ticking time bomb' approach, 26
tools
    commercial security testing, 53–55
    hacking and cracking, 44–47
    packet sniffing, 42–44
    reverse engineering, 47–52
    security scanners, 38–44
    for testing buffer overflow vulnerabilities, 129–130
    Web site test, 47
topographical attacks, 330
traffic, recording with sniffers, 42–43
Transmission Control Protocol (TCP)
    described, 32
transport layer in multilayer model, 32
Trojan DLLs, 101
Trojan horses, 22

U

UDP (Universal Datagram Protocol)
    described, 33
unauthorized access defined, 23–24
Unix
    crypt () function, 45
    Nessus security scanner, 39–40
    password hashing functions, 197–198
    permission mode bits (table), 63
    timesharing, user categories, 62–64
    writing main crack routine, 82–83
Unix Network Programming (Stevens), 27
URLs (universal resource locators)
    disguised, 336–338
    forceful browsing and, 280
    lists of safe, 282
    spoofing in the browser, 329–330
    uncommon schemes, 330–331
usernames
    and password-cracking tools, 45
    unauthorized use, 258–259
    Web forms' use of, 297–301
users
    authentication, vs. Web server authentication, 295
    categories in Unix permission scheme, 62–64
    passwords. *See* passwords

V

validating input, 7–8, 273
Valleri, Marco, 43
values, cookie, 301–302
variables, initializing, 211
VeriSign, Authenticode signing capability, 101
Verity Ultraseek, 264
Viega, John, 17
virtual stores, 223
VirtualLock, VirtualUnlock () API functions, 230
viruses
    described, 22
    'I Love You', 88, 92, 94
    Melissa, 88
Visual Basic and Melissa virus, 88
volunteering excess information, 255–263
vulnerabilities
    See also specific vulnerability
    ACL, superuser models, 64
    buffer overflow. *See* buffer overflow vulnerabilities
    BugTraq mailing list of, 338
    cross-site scripting, 267–275
    described, 21
    dtappgather, 67
    dynamic linking and loading, 95–103
    forceful browsing, 277–283
    format string, 158, 158–170
    HTTP header tampering, 306–307
    integer overflow, 173–181
    learning from, 335–338
    memory, leaving things in, 211–221
    passwords in plaintext, 187–196
    query strings, 306
    reporting, 26
    SQL injection, 311–319
    SQLsnake, 75
    swap files and incomplete deletes, 223–230
    volunteering excess information, 255–263
vulnerability databases, online, 12

W

warning users of potential malicious action, 92
weak passwords
    finding, fixing, 75–86
    shipping products with, 83
    vulnerabilities of, 73–75
Web applications
    See also applications
    client and browser data manipulation, 297–301
    insecure, and SQL injections, 256
    preventing forceful browsing in, 295
Web browsers
    and forceful browsing, 278–283
    Internet Explorer. *See* Internet Explorer
    programs described, 277
    security issues, 325–328
    spoofing of URLs, 329–331
    and unsafe ActiveX controls, 328–329
Web forms, client data access, 297–301
Web pages, SSIs in, 90
Web servers and de-escalation, 70
Web sites
    Authenticode, 101
    cross-site scripting vulnerability, fixing, 267–275, 276
    exploiting through SQL injection, 316–319
    finding vulnerabilities, 307–308
    Internet Engineering Task Force RFC documents, 34
    online vulnerability databases, 12
    Pablo Software Solutions, 258
    test tools, 47
    useful tools, 55
WebProxy Web testing tool, 53
WEP (Wired Equivalent Privacy), 252–253
white-box testing for buffer overflow vulnerabilities, 128
'white' data, 93
Windows
    API monitoring in, 50
    bitmap format, 135–136
    and Code Red worm, 127
    deleting page files upon shutdown, 231
    and escalation of privilege vulnerabilities, 68–71
    finding temporary files vulnerabilities, 206
    hashing function, 81
    OllyDbg debugger for, 51
    passwords, cracking, 77
    searching memory with custom debugger, 215–220
Windows 98 credential crashing, 214
Windows Cryptographic API, hashing support, 198
Windows for Workgroups, 214
Windows Update, 40
WinHex editor, 49, 116, 227
WinMySQLadmin, 188
WinPcap, 43
Winsock controls and SimpleFuzz, 142
Wired Equivalent Privacy (WEP), 252–253
Word, duplicating cells in (fig.), 180
worms
    Code Red, 127, 132
    Morris Internet, 86
write permissions in Unix permissions scheme, 63
Writing Secure Code (Howard and LeBlanc), 6, 17
Wysopal, Chris, 45

X

XML format, 148

Y

Yu, Liu Die, 332

Z

ZeroMemory () function, 221